Eduqas
Physics
A Level

Revision Workbook 1

Gareth Kelly

Iestyn Morris

Nigel Wood

Published in 2021 by Illuminate Publishing Limited, an imprint of Hodder Education, an Hachette UK Company, Carmelite House, 50 Victoria Embankment, London EC4Y 0DZ

Orders: Please visit www.illuminatepublishing.com or email sales@illuminatepublishing.com

British Library Cataloguing in Publication Data

A catalogue record for this book is available from the British Library

ISBN 978-1-912820-63-4

Printed by Severn, Gloucester

08.21

The publisher's policy is to use papers that are natural, renewable and recyclable products made from wood grown in sustainable forests. The logging and manufacturing processes are expected to conform to the environmental regulations of the country of origin.

Publisher: Eve Thould

Editor: Geoff Tuttle

Design and layout: Nigel Harriss

Cover image: © NASA

Acknowledgements

The authors would like to thank Adrian Moss for guiding us through the early stages of production and the Illuminate team of Eve Thould, Geoff Tuttle and Nigel Harriss for their patience and meticulous attention to detail. We are also indebted to Helen Payne for her eagle eyes in spotting mistakes and inconsistencies and for her many insightful suggestions.

Contents

Introduction

How to use this book   4

Assessment objectives   5

Preparing for the examination   7

Common exam mistakes   10

PRACTICE QUESTIONS

Component 1: Newtonian Physics

Section 1: Basic physics   11

Section 2: Kinematics   23

Section 3: Dynamics   35

Section 4: Energy concepts   45

Section 5: Circular motion   55

Section 6: Vibrations   63

Section 7: Kinetic theory   73

Section 8: Thermal physics   81

Component 2: Electricity and the Universe

Section 1: Conduction of electricity   91

Section 2: Resistance   97

Section 3: DC circuits   107

Section 4: Capacitance   119

Section 5: Solids under stress   129

Section 6: Electrostatic and gravitational fields of force   139

Section 7: Using radiation to investigate stars   148

Section 8: Orbits and the wider universe   156

PRACTICE EXAM PAPERS

Component 1: Newtonian Physics   168

Component 2: Electricity and the Universe   181

ANSWERS

Component 1 Practice questions   196

Component 2 Practice questions   215

Component 1 Practice paper   231

Component 2 Practice paper   234

How to use this book

What this book contains

The A level physics course consists of three components, each assessed with a separate exam. This book covers the first two components.

Component 1: Newtonian Physics   **Component 2: Electricity and the Universe**

Component 1 is divided into eight areas of study; Component 2 also has eight areas of study.

Each area of study has its own section in this book and there you will find:

- A **concept map**, which displays how the different concepts within the area of study are related to one another and to other areas of study.
- A set of **graded questions**, with space for your answers, which are designed to test the content of the area of study in a way that is similar to the examination.
- A **question and answers** section containing one or two examples of exam-style questions, with answers from two students, Rhodri and Ffion (who produce answers of different standard), together with examiner marks and discussion.

The next section comprises two **practice papers** – one for each component. The final section consists of **model answers** to all the graded questions and the practice papers.

How to use this book effectively

This book can be used exclusively for revision, in which case you will work your way gradually through the questions as you revise each area of study. Alternatively, it can be used regularly as you are learning A level physics, with each set of questions providing an end-of-area check/test. Your teacher might even like to use it as a homework book because it contains 16 sets of end-of-area questions as well as the practice papers.

Probably the worst method of revising physics is just reading your notes or your textbooks, a technique that will just send you to sleep. Regardless of the quality of writing of the notes/textbooks, there is an inherent, soporific effect to reading notes. One way of combating this tendency to lose concentration is to make your own notes as you read through the notes/textbook. Even so, you will be preparing yourself only for exam questions in which you reproduce learned material. These, so-called Assessment Objective 1 (AO1) questions, form only 30% of the exams. The higher-level skills required for AO2 and AO3 require different revision techniques. See pages 5 to 8 for further information on assessment objectives.

You will find that the best way of revising is to answer exam-type questions. As in many fields, it is practice that makes perfect in physics and you should practise as many past papers and questions within this book as you can.

You will, inevitably, come across some questions that seem difficult. Should you not be able to answer them, it is time to visit your notes/textbook, but this time you will be doing so with a purpose to help you concentrate. Should the question still leave you stumped, have a look at the answers in the back of the book. If the answer seems unintelligible, it is time to ask your teacher or your fellow students for an explanation of how and why the answer is correct. Your teacher may also point out that different answers are creditworthy, especially in discussion questions.

So, answering the questions and analysing the model answers will do more for your exam preparation than just reading notes or making revision timetables (however colourful)!

Assessment objectives

You need to demonstrate expertise in answering examination questions in different ways. Because it's physics, you will need to use maths in most questions. Some questions require expertise in practical skills, even in the written papers. However, besides these two categories there are the assessment objectives. There are three of them:

Assessment Objective 1 (AO1)

AO1 questions are ones in which you need to:

demonstrate knowledge and understanding of scientific ideas, processes, techniques and procedures

This objective accounts for 30% of the marks in the **three** component papers.

The official description in bold (**demonstrate … procedures**) sounds far more complicated than is necessary. Essentially, these are the marks that can be obtained without too much thinking. This category covers:

- Recall of definitions, laws and explanations from the specification
- Inserting appropriate data into equations
- Deriving equations where required in the specification
- Describing experiments from the **specified practicals**.

In general, no judgement or new thinking is required. One can learn a definition or a law without fully understanding it, so it is a good idea to use the Eduqas terms and definitions (T&D) booklet to help you memorise these. Defined T&D from the booklet are printed in **bold** in the concept map for each area of study in this book.

Examples of AO1 questions

1. State the conditions for a body to remain in equilibrium. [2]

Good answer (full marks 2/2): Resultant force = 0, resultant moment about any point = 0.

Bad answer (no marks 0/2): The up and down forces must cancel and the clockwise moment = anticlockwise moment.

The *bad answer* is not general enough – there are other possible directions for forces, and the moment explanation assumes that there is only one clockwise moment and one anticlockwise moment.

2. Define internal energy. [2]

Good answer (2/2): It's the sum of the random kinetic and potential energies of the molecules.

Bad answer (0/2): For an ideal gas it's the KE of the particles.

The bad answer is missing 'sum of' and potential energy. Note that the abbreviation KE would be perfectly acceptable and that the good answer even includes the word 'random'.

3. Calculate the photon energy of e-m radiation of wavelength 3.0×10^{-10} m. [2]

Good answer (2/2): $E = hf = \dfrac{hc}{\lambda} = \dfrac{6.63 \times 10^{-34}\,\text{J s} \times 3.00 \times 10^{8}\,\text{m s}^{-1}}{3.00 \times 10^{-10}\,\text{m}} = 6.63 \times 10^{-16}$ J

Bad answer (0/2): $E = hf = 6.63 \times 10^{-34} \times 3.00 \times 10^{-10} = 1.989 \times 10^{-43}$ J

Although the *bad answer* used a correct equation ($E = hf$), the student didn't insert the data correctly (the wavelength was inserted instead of the frequency) and so gained no credit. Putting the units in the working might have made it easier for the student to spot the mistake. Note that this question seems like Component 3 but would be expected for spectra (2.7).

Assessment Objective 2 (AO2)

AO2 questions are ones in which you need to:

> **apply knowledge and understanding of scientific ideas, processes, techniques and procedures:**
> 1. **in a theoretical context**
> 2. **in a practical context**
> 3. **when handling qualitative data**
> 4. **when handling quantitative data.**

Here, the key words are 'apply knowledge'. The application of knowledge is required in theoretical, practical, qualitative and quantitative contexts. Theoretical just means some idealised context made up by the examiner. Practical means that the data have apparently come from a real experiment (although the data are usually made up by the examiner). *Qualitative* means without numbers and calculations whereas *quantitative* means the opposite (i.e. with numbers and calculations).

Note that application of knowledge here can also include analysis of data even though 'analyse' also appears in AO3. Generally, if you are told what type of analysis to carry out, these will be AO2 skills. If the question is more open-ended and you must choose the analysis methods yourself, the question will be classified as AO3. Questions relating to AO2 are the most common type and account for 45% of the marks on the papers. Note that all calculations must be mainly AO2 marks: we have seen that inserting data into an equation is classed as AO1, but any manipulation of an equation, such as changing the subject, and the production of a final answer is AO2 (unless it is AO3).

Examples of AO2 questions

1. A force of 6.96 N places a wire of cross-section 4.3×10^{-5} cm^2 under a strain of 2.6%. Calculate its Young modulus. [3]

   Good answer (3/3): $E = \dfrac{\sigma}{\varepsilon} = \dfrac{6.96\,\text{N}/4.3 \times 10^{-9}\,\text{m}^2}{0.026} = 6.2 \times 10^{10}$ Pa

   Bad answer (1/3): $E = \dfrac{\sigma}{\varepsilon} = \dfrac{6.96/4.3 \times 10^{-5}}{2.6} = 6.2 \times 10^{4}$ Pa

   The two mistakes here are not to convert the cm^2 to m^2 and to express the 2.6% as 0.026.

2. Calculate the rms speed of neon atoms of mass 3.32×10^{-26} kg at 350 K. [2]

   Good answer (2/2): $\frac{1}{2}mc^2 = \frac{3}{2}kT \longrightarrow \sqrt{\overline{c^2}} = \sqrt{\dfrac{3kT}{m}} = \sqrt{\dfrac{3 \times 1.38 \times 10^{-23} \times 350}{3.32 \times 10^{-26}}}$ m s^{-1}

   Bad answer (0/2): $\frac{1}{2}mc^2 = \frac{3}{2}kT \longrightarrow \sqrt{\overline{c^2}} = \dfrac{3kT}{m}$ m s^{-1}

   The bad answer has not done enough to obtain the 1st mark (substituting into a correct, relevant equation) – forgetting to do the square root has cost 2 marks.

3. Calculate the distance between two stars of mass 1.3×10^{30} kg and 1.9×10^{30} kg whose period of mutual orbit is 12.3 days. [3]

   Good answer (3/3): Rearranging $T = 2\pi\sqrt{\dfrac{d^3}{G(M_1 + M_2)}}$; $d = \sqrt[3]{\dfrac{T^2 G(M_1 + M_2)}{4\pi^2}} = 1.83 \times 10^{10}$ m

   Bad answer (0/3): $27.3 = 2\pi\sqrt{\dfrac{d^3}{9.81 \times (3.2 \times 10^{30})}}$ (time in day, no algebra, G wrong $\longrightarrow$ 0).

Assessment Objective 3 (AO3)

AO3 questions are ones in which you need to:

> **analyse, interpret and evaluate scientific information, ideas and evidence, including in relation to issues, to:**
> 1. **make judgements and reach conclusions**
> 2. **develop and refine practical design and procedures.**

These questions account for 25% of the marks in the three component papers.

The verbs *analyse*, *interpret* and *evaluate* are all appropriate as these are, indeed, what you will have to do. Most of these AO3 marks will concentrate on the first point – *judgements* and *conclusions*. The context will often be similar to one of the specified practicals with realistic data. Your analysis may well include analysing graphs to make numerical conclusions. You might have to evaluate the quality of the data and your conclusions. In some questions you are given a statement and have to determine whether or not (or to what extent) it is true. There are usually several ways of getting a sensible answer: you must choose one and structure your answer carefully. Other questions relate to the second part of the AO3 statement – develop and refine practical design and procedures. Usually, these questions are based on imperfections in the data and how you could improve the procedure or the apparatus to obtain better data. To answer these questions, you will need to read them carefully because there will be a clue (perhaps right at the start) as to what went wrong.

Another type of question is based on the part of the statement *including in relation to issues'*. The **'issues'** include: risks and benefits; ethical issues; how new knowledge is validated; how science informs decision making. Try and make sensible comments; the mark scheme will allow for many approaches and the marks will be quite attainable – approach the questions like a politician: have a view. Every theory paper has one issues question.

Examples of AO3 questions

1. Discuss the quality of data obtained by Gwynfor for obtaining the Young modulus of copper and whether, or not, the data are in good agreement with theory. [4]

Good answer (4/4): The data appear good because the line of best fit passes through all error bars. The best fit line starts off as a straight line through the origin, in agreement with theory, the gradient then decreases as the data passes the elastic limit of copper, also in agreement with theory. The final value for the Young modulus of copper is 25% too low and is inaccurate (since the answer to the previous part suggested an uncertainty of only 15%).

Bad answer (0/4): I like the data because there is a definite pattern to the results and the graph is the same shape as the one that appears in the textbook. The final value of the copper YM seems too low and therefore is inaccurate.

2. The value of the acceleration due to gravity obtained in this pendulum experiment is inaccurate because it is too high ($10.1 \text{ m s}^{-2} \pm 0.2 \text{ m s}^{-2}$). Suggest a reason for this inaccuracy (see diagram of apparatus). [2]

Discussion: This is an AO3 question and an example of the 'suggest' command. The string in the diagram might have been labelled 'thick string' and you might be expected to mention that the centre of mass of the string is halfway along its length. The true length of the 'pendulum' is therefore shorter, giving a shorter period and a greater value for g (tough!).

Preparing for the examinations

Exam mark schemes and examiner comments

When examiners mark examination papers, they do so using agreed mark schemes. This means that different examiners use the same criteria when marking. These mark schemes are not designed for students. It will take a bit of work but it is worth looking at the mark schemes in the **Question & Answer** sections; they will give you some hints on what you need to do to get the best marks. This is a section of the mark scheme for the Q&A on p20.

Question part			Description	AOs			Total	Skills	
				1	2	3		M	P
(b)	(i)	(II)	Correct equation from applying Principle of moments, or by implication ecf on 11.8 N and 35.3 N [1] 2nd correct eqn, e.g. $T_A + T_B = 11.8 + 35.3$ or PoM about a 2nd point. [1] $T_A = 14.0$ N (accept 14 N) [1] $T_B = 38.1$ N (accept 38 N) [1] Ecf on second force to be calculated if wrong only because of mistake in first.		4		4	4	

The 'description' says what the examiner is looking for.

First mark: Any correct Principle of Moments (PoM) equation from this set-up will get you this. Ignore 'by implication' for now: ecf stands for *error carried forward* and it means that, if you got the wrong answers in part (I), which is where the 11.8 N and 35.3 N come from, you can get full marks if you apply these wrong answers here.

Second mark: The second equation could be a second PoM equation, about a different point, or (easier) equating the upward and downward forces on the rod.

Third and fourth marks: for the answers. Note that ecf still applies on wrong answers in part (I), (examiners always have their calculators out). Also, if you get T_A wrong but use this answer correctly to find T_B, that will gain the mark ecf.

What does 'or by implication' mean? This means that, if you haven't written this down but have obtained the correct answers, the examiner will assume that you've also done this.

Don't worry about the M and P columns: these just tally up the number of marks that count towards the required Maths and Practical marks totals for the paper.

Examiner's annotations:

When marking, an examiner uses a ✓ where a mark is earned and (often) a ✗ where a mark has been missed. They will also write ✓ ecf where they are awarding a mark on this basis. Other annotations: 'bod' stands for 'benefit of the doubt'; 'not enough' indicates that the answer is not wrong but that it isn't complete enough for the mark. The Component 1 and 2 papers are 100 marks and last 135 and 120 minutes respectively.

Component 1 – 100-mark paper of duration 135 minutes

Section A: 80 marks. Containing eight areas of study, it will have a mean of 10 marks per area of study and you might expect eight questions – one on each area of study. While each year's paper will differ significantly from this basic structure, the examiners will try to distribute their marks equitably between the areas of study, and eight questions will be a reasonable rule of thumb for Section A.

Section B: Comprehension – This is usually a two-page passage based loosely on A level physics and always valued at exactly 20 marks. Previous topics have included astronomy, cosmology, particle physics, mobile phone experiments, rocket experiments – you will not know what to expect. Nonetheless, the questions will nearly all be answerable using only A level physics. If there are any questions not based on A level physics, then that physics should be in the passage and presented at an understandable level. It is difficult to prepare for the comprehension questions but practise previous examples and be confident in your physics knowledge and you will be fine.

Additionally, there are four things (other than randomness of distribution) that arise to mess up this beautifully symmetric system:

1. Practical content: You can expect 20% of the examination (20 marks) to be based on experimental analysis. This usually means that one (or two) of the questions will be based on one of the eight specified practicals for this component. This could be a description of the method, error analysis, graphs and conclusions – often the longest question on the paper.

2. Quality of extended response (QER): This is a 6-mark question with a lot of lines for writing and maybe some space for diagrams, too. These tend to be AO1 marks and so rely on you learning the basic physics required to answer the question. This, however, is only part of the problem. Not only must you put the required information down on paper but you must also do so in a logical, well-presented format, employing good language skills. The penalty for poor English is generally only 1 or 2 marks but the penalty for not knowing the relevant physics is 6 marks! Describing a specified practical is a common QER question.

3. Synoptic content: Although the three component exams usually follow in order a few days apart, you still need to ensure that you have revised Components 2 and 3 thoroughly before the Component 1 exam because of this synoptic content. Any of the areas of study of Components 2 and 3 can be combined with a Component 1 area of study to make a more difficult question, e.g. circular motion might be combined with gravitational motion.

4. Issues: See the earlier AO3 section.

Component 2 – 100-mark paper of duration 120 minutes

This component also contains eight areas of study leading to a mean of 12½ marks per area of study and the rule of thumb this time will still be eight questions. Note that, sometimes, two areas of study might be combined into one longer question and, likewise one area of study might be split into two smaller questions. Everything about practical content, QER, synopticity and issues applies equally to Component 2 (but it has no Section B).

Practical content in Components 1, 2 (& 3): Examiners will make every effort to ensure that the practical skills examined in Components 1, 2 (& 3) are as widespread as possible, e.g. plotting points will not be asked for in all three components. The same goes for the other practical skills, such as measuring gradients, describing lines of best fit. So, after Component 1 you'll probably know what **not** to expect in Component 2 (and after Components 1 and 2 you'll have a reasonable guess at the skills required in Component 3).

Key command words and phrases in examination questions

These are the words or phrases which let you know what sort of answer is expected – there are quite a few to look out for.

State: Just provide a statement without an explanation.

> Example: State what happens to the current as the temperature of the metal wire increases.

> Answer: It decreases.

Define: You need to provide a statement which is close to (or equivalent to) that which appears in the Eduqas Terms and Definitions booklet.

> Example: Define the potential difference (pd) between two points.

> Answer: It's the energy converted from electrical potential energy (to other forms) per unit charge (passing between the points).

Explain what is meant by (or explain the meaning of ...): This can mean a couple of things:

1. Sometimes it just means the same as 'define', e.g. explain what is meant by the potential difference (pd) between two points.
    Answer: (Exactly the same as above).

2. Sometimes it's a definition with a number included, e.g. explain what is meant by the statement 'The Young modulus of steel is 2×10^{11} Pa'.
    Answer: This is the stress divided by the strain **and** for steel it is 2×10^{11} Pa.

Explain the difference (between two things): This is two definitions in disguise because if you define both things you have automatically explained the difference between them.

> Example: Explain the difference between a vector and a scalar.

> Answer: A vector has magnitude and direction whereas a scalar only has magnitude.

Describe: Provide a brief description but no explanation is required.

> Example: Describe the appearance of an emission spectrum.

> Answer: Bright lines on a dark background.

Explain ... (some statement): Sometimes this requires a logical argument.

> Example: Explain briefly how an exo-planet causes a star to 'wobble'.

> Answer: The star and planet orbit their centre of mass. Hence, it is the star's orbit that produces the 'wobble' (usually detected by Doppler shifts of absorption lines).

Suggest ... (or suggest a reason ...): Although not a common command word, this can produce some questions that are difficult to answer. These will often be AO3 marks, appearing at the end of a question requiring evaluation skills.

> Example: Suggest a reason why the gradient of the temperature against time curve decreases at higher temperature.

> Answer: The gradient decreases because more heat is being lost as the temperature difference between the aluminium block and the air increases (hence, more energy input is required for each unit temperature rise).

Calculate or **determine**: The aim is to obtain the correct answer (along with the correct unit if required by the mark scheme). With this command word, the correct answer will obtain full marks without the workings. However, you are advised strongly to show your working as marks are available for this even if the answer is wrong.

> Example: Calculate the mass of a 2.00 cm diameter steel ball of density 7800 kg m^{-3}.

> Answer: $m = \rho V = \rho \frac{4}{3}\pi r^3 = 7800 \text{ kg m}^{-3} \times \frac{4}{3}\pi \times (0.0100 \text{ m})^3 = 0.033 \text{ kg}$ (2 sf)

> [Note that you do not have to put units in the calculation – but you do in your answer!]

Compare: Not a common command word but you ought to do what it says on the tin – compare the things it says to compare in the question.

> Example: Compare the appearance of a hot star (10 000 K) with that of a cool star (3000 K) with the same diameter.
>
> Answer: The hot star will appear brighter and slightly blue, and the cool star will appear red.

Evaluate: You will be required to make a judgement, e.g. whether a statement is correct or wrong, or to decide whether data are good or a final value is accurate.

Justify: This is sometimes used in a very similar manner to the word 'determine' when AO3 marks are being examined, e.g. justify Jay's statement that the 2.0 V reading was anomalous.

Discuss: This can often be a command word in the 'issues' question. In general, you will not go far wrong if you make a couple of points in favour of the discussion issue, a couple against it, and then draw some sort of a sensible conclusion.

Common exam mistakes

1. **Not converting the given numbers correctly**. Planetary distances are usually in km while wire radii are in mm. Resistors can be in W, kW, MW and these values have to be converted to the correct powers of 10. There are other common conversions such as changing diameter to radius when using area or volume formulae. Failures here are simple mistakes and do not show a poor understanding of physics but are penalised one mark most of the time.

2. **Not reading the question carefully enough**. This usually results in not answering the question that was asked – either by answering a different question altogether or by missing part of the question. The most common parts of questions that are omitted are those that do not have dotted lines for you to answer on, e.g. adding to diagrams. Pay particular attention to these short parts of questions. Other common missed questions are ones that have an **and** condition in the question itself, e.g. calculate the magnitude **and** the direction. One or other part of the question will have been forgotten in the answer.

3. **Not understanding equations properly**. This often involves substituting wrong values into equations – an unforgivable sin! In kinematics equations, for instance, u and v are often mixed up. You shouldn't really have to use the data booklet; you should know the equations intimately and only check it from time to time to ensure that you recollect them correctly. How do you ensure that you don't misunderstand an equation? Practise, practise, practise!

4. **Not knowing the basic Terms and Definitions** (a common cause of loss of marks). There is an Eduqas booklet full of these – you should know everything within its covers.

5. **Forgetting to square a value in the equation**. This happens most often with the kinetic energy equation – the equation $E = \frac{1}{2}mv^2$ is written correctly but then the candidate forgets to square the velocity on the calculator. Or the converse: forgetting to square root the answer when using the same equation to calculate the velocity!

6. **Not planning the structure before answering the QER** (and extended explanations). Too many QER responses are rambling and unstructured. This is easily remedied by spending a moment to plan and structure your answer. Using short sentences tends to help too.

7. **Not matching the correct corresponding values in a calculation**. By far the most common mistake here is with electrical circuits: current, pd and resistance, e.g. a pd and a current will be combined to obtain a resistance ($R = V / I$) but the current and pd do not match – the pd is for one resistor and the current for another.

8. **Bad algebra**, i.e. rearranging equations incorrectly. This is a common and very expensive mistake. If an equation is rearranged incorrectly in a calculation you will be extremely fortunate if the mark scheme allows you even one mark. Most often, the simple equations are rearranged incorrectly, e.g. $R = \frac{I}{V}$ No! $I = \frac{R}{V}$ No! $m = \frac{\rho}{V}$ No! $a = \frac{m}{F}$ No! No! No!

Component 1: Newtonian Physics

Section 1 Basic physics

Topic summary

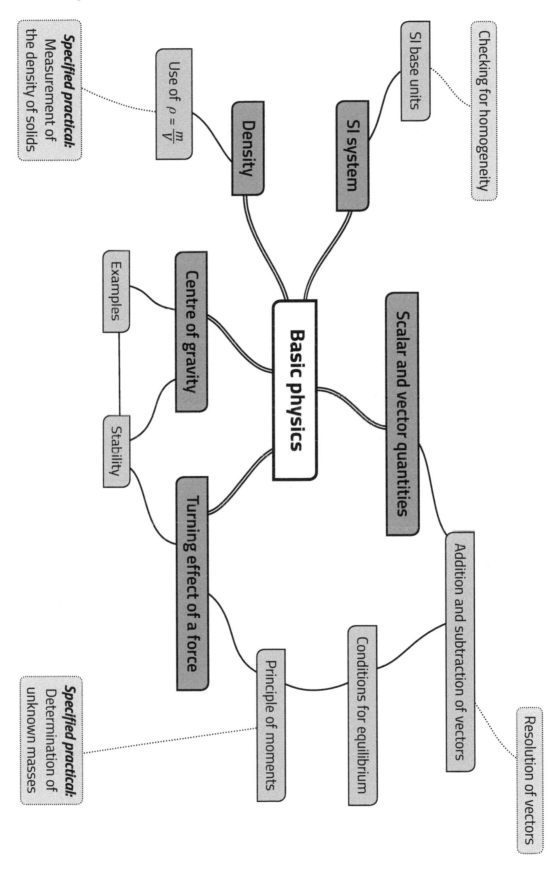

Specified practical: Measurement of the density of solids

Use of $\rho = \frac{m}{V}$

Density

SI base units

Checking for homogeneity

SI system

Examples

Centre of gravity

Stability

Basic physics

Scalar and vector quantities

Turning effect of a force

Addition and subtraction of vectors

Principle of moments

Conditions for equilibrium

Resolution of vectors

Specified practical: Determination of unknown masses

Q1 The candela (cd) is one of the seven SI base units. It is not used in A level physics. State the names and symbols of the other six SI base units. [2]

....

....

....

Q2 Newton's second law of motion can be written $F = ma$.

(a) State the SI unit of force and give its symbol. [1]

....

(b) Use the equation to write the unit of force in terms of the metre (m), kilogram (kg) and second (s). [2]

....

....

....

(c) Use the answer to (b) and the defining equation for work to show that the joule, J, can be expressed as $kg\, m^2\, s^{-2}$. [2]

....

....

....

Q3 Very slow-moving objects in air experience a drag force, F, proportional to their speed, v, through the air, that is $F = kv$, where k is a constant.

(a) Show that the unit of k can be written $[k] = kg\, s^{-1}$. [2]

....

....

....

(b) For more rapidly moving objects, the drag equation becomes $F = KAv^2$, where A is the cross-sectional area and K is a constant (different from that in part (a)). Express the unit of K in terms of the base SI units. [2]

....

....

....

Q4 The area, A, of a circle is related to its radius by the equation $A = \pi r^2$. Explain why π has no unit. [1]

....

....

Q5 The following is a list of quantities met in A level physics. Divide them into *scalar* and *vector* quantities. [2]

energy   acceleration   time   density   temperature   velocity   momentum   pressure

....

Q6 One of the kinematic equations for constant acceleration is $v = u + at$. Show that this equation is homogeneous. [2]

....

....

....

Q7 Two forces act on an object. The forces have magnitudes of 28 N and 45 N. Draw diagrams to show how the resultant force can have magnitudes of: (a) 73 N, (b) 17 N and (c) 53 N. [4]

In each case, state the direction of the resultant force relative to the 45 N force.

[Hint: $28^2 + 45^2 = 53^2$]

(a)

Direction = ...

(b)

Direction = ...

(c)

Direction = ...

Q8 A force acts at 25° to the horizontal. The magnitude of the vertical component of the force is 53 N. Calculate:

(a) The magnitude of the force. [2]

....

....

....

(b) Its horizontal component. [1]

....

....

Q9 Two tugs, T_1 and T_2, are towing a large ship out of port. T_1 exerts a force of 5.0 kN at 15° to the forward direction. T_2 is a more powerful tug and exerts a larger force F at 10° to the forward direction. The resultant force of these is exactly in the forward direction.

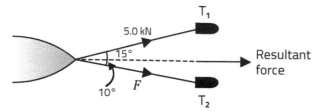

(a) Show that the magnitude of the component of the 5.0 kN force at 90° to the forward direction is approximately 1.3 kN. [2]

....

....

....

(b) State the magnitude of the component of force F at 90° to the forward direction. [1]

....

(c) Calculate the magnitude of force F. [2]

....

....

....

(d) Calculate the resultant of the two forces on the ship. [3]

....

....

....

....

....

Q10 An object has several forces acting on it. State the two conditions which must be satisfied for the object to be in equilibrium. [2]

....

....

....

....

....

Q11 A ball is thrown through the air. At time t_1 it is travelling with a velocity of 15.0 m s^{-1} at 30.0° to the horizontal and climbing. At time t_2 its velocity is 20.0 m s^{-1} at 49.5° to the horizontal and falling.

(a) Show that its horizontal and vertical components of velocity at t_1 are 13.0 m s^{-1} and 7.50 m s^{-1} upwards, respectively. [2]

....

....

....

(b) By considering the horizontal and vertical components of velocity at t_2 and your answers to part (a), calculate the change in velocity of the ball between t_1 and t_2. [3]

....

....

....

....

....

Q12 A ball, travelling at 12 m s^{-1}, collides with a wall and rebounds at 10 m s^{-1} at right angles to its original direction. The diagram shows this seen from above.

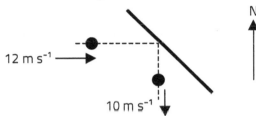

Calculate the magnitude and direction of the change in velocity. [3]

....

....

....

....

Q13 A spring has a spring constant k. This is the force per unit extension needed to stretch the spring and has the unit N m^{-1}. If an object of mass m is hung from the spring, pulled downwards and released, it oscillates with a period, T. A student cannot remember whether the correct equation for calculating T is:

$$T = 4\pi^2 \frac{m}{k}, \qquad T = 4\pi^2 \frac{k}{m}, \qquad T^2 = 4\pi^2 \frac{m}{k}, \qquad \text{or} \quad T^2 = 4\pi^2 \frac{k}{m}$$

By considering the units of T, m and k, evaluate whether any of these equations could be correct. [3]

....

....

....

....

....

Q14 A reel contains copper wire with a diameter of 0.317 mm. Calculate:

(a) The cross-sectional area of the wire in m^2. Give your answer in standard form. [2]

....

....

(b) The volume of a 85.0 cm length of the wire. [2]

....

....

(c) The length of this wire that could be produced from a block of copper of mass 2.50 kg. [The density of copper is 8.96×10^3 kg m^{-3}.] [3]

....

....

....

....

....

....

Q15 (a) A gas at a pressure of 2.50 MPa is held by a piston in a cylinder of radius 10.0 cm. Calculate the force that the gas exerts on the piston. [2]

....

....

....

(b) A sample of gas is held in an open-ended pipe by a piston consisting of a solid disc of copper, of length 10.0 cm, as shown. The pressure, p_A, of the air above the copper disc is 101 kPa. Calculate the pressure, p, of the gas. [Hint: the disc is held in position by a net upward force due to the pressure difference.] [4]

(Density of copper = 8.96×10^3 kg m^{-3})

....

....

....

....

....

....

....

Component 1 Practice questions

Q16 Rhian is given a rectangular block of iron and asked to determine its density. These are her measurements:

length / cm	6.35, 6.38, 6.34, 6.38, 6.37
width / cm	4.26, 4.24, 4.28, 4.17, 4.25
height / cm	2.79, 2.81, 2.83, 2.80, 2.81
mass / g	599.5

(a) Rhian thinks that she made a mistake with one of her measurements, so decides to ignore it. Identify the suspect measurement and give a reason. [2]

(b) In calculating the uncertainty in her value for the density of the iron, Rhian decides to ignore the uncertainty in the mass measurement. Evaluate her decision. [2]

(c) Use the data to calculate a value for the density of iron together with its **absolute** uncertainty. [5]

Q17 Maurice is given a uniform **half-metre** rule, a 100 g mass and a piece of cord and asked to measure the mass, m, of a piece of metal. He suspends the 100 g mass from the 1.0 cm mark and balances the ruler on the edge of a pencil. The balance point is 15.0 cm.

pencil

He replaces the 100 g mass with the piece of metal. The balance point is now 12.5 cm.

Use these results to determine the mass of the piece of metal. [4]

Q18 A uniform public house sign of width 80 cm and mass 3.5 kg is suspended from a metal bar of mass 1.5 kg and length 90 cm by two links, **A** and **B**, each of which is 10 cm from the edge of the sign. The metal bar is attached to the wall by a hinge, **H**. Link **A** is 15 cm from **H**. The metal bar is also attached to the wall by a wire which is attached to the bar at **B**.

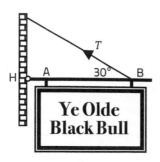

(a) Explain why the tensions in the two links, **A** and **B**, are equal and calculate their magnitude. [3]

....

....

....

....

....

(b) Calculate the sum of the clockwise moments, about **H**, of the forces on the bar. [2]

....

....

....

....

(c) By considering moments about **H**, calculate the tension, T, in the wire, stating which principle for equilibrium you use. [3]

....

....

....

....

(d) The hinge, **H**, also exerts a force on the metal bar. Using a different condition for equilibrium, determine the magnitude and direction of the force exerted by the hinge on the bar. [3]

....

....

....

....

....

Q19 Dominic completes a table of vector and scalar values:

Vectors	Scalars
Force	Temperature
Work	Displacement
Velocity	Energy

(a) State which of Dominic's choices are incorrect. [2]

....

....

....

(b) In fluid dynamics, the viscosity (μ) of a fluid can be given by the equation:

$$\mu = \frac{\rho u L}{k}$$

where k is a dimensionless number, ρ the density of the fluid, u the speed of the flow and L is a characteristic length.

(c) (i) Explain why N s m^{-2} is a valid unit for the viscosity, μ. [4]

....

....

....

....

....

(ii) The dimensionless number, k, in the equation is an important factor in determining the motion of a cricket ball. Calculate k for a cricket ball with characteristic length (L) of 7.1 cm, travelling through air of density 1.16 kg m^{-3} and viscosity 1.87×10^{-5} N s m^{-2} at a speed of 41.2 m s^{-1}. [3]

....

....

....

....

....

Question and mock answer analysis

Q&A 1

(a) An object is acted on by forces whose lines of action are all in one plane. State the conditions necessary for the object to be in equilibrium. [3]

(b) In a theatre a spotlight of mass 3.60 kg hangs from a uniform rod of length 3.00 m and mass 1.20 kg. The rod is kept horizontal by vertical wires attached to points A and B as shown.

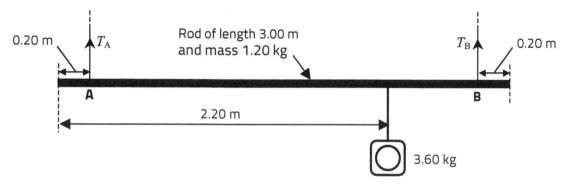

(i) (I) Use labelled arrows to show the magnitude and direction of the two other main forces on the system. [2]

(II) Determine the tensions T_A and T_B. [4]

(ii) Explaining your reasoning, determine the least mass that can be hung from *anywhere* on the bar to make the bar tilt. The 3.60 kg spotlight has been removed. [3]

What is being asked

This question is about the equilibrium of objects when acted upon by a number of forces. It has several parts involving AO1, AO2 and AO3, with verbal and mathematical skills. It starts with a recall part (a), on the conditions for equilibrium of an object under the influence of a number of forces. This part is designed to point you in the right direction for the analysis and evaluation of part (b).

Mark scheme

Question part			Description	AOs			Total	Skills	
				1	2	3		M	P
(a)			1. The [vector] sum of the forces on the object is zero or equivalent. [1] 2. The sum of the clockwise moments *about any point* equals the sum of the anticlockwise moments *about that point.* Non italic: [1], italic: [1]	3			3		
(b)	(i)	(I)	Vertical downward arrow from centre of rod, labelled 11.8 N (unit needed)[1] Vertical downward arrow on the spotlight, labelled 35.3 N (unit needed) [1]		2		2	2	
		(II)	Correct equation from applying Principle of moments, **or** by implication ecf on 11.8 N and 35.3 N [1] 2nd correct eqn, e.g. $T_A + T_B = 11.8 + 35.3$ or PoM about a 2nd point. [1] $T_A = 14.0$ N (accept 14 N) [1] $T_B = 33.1$ N (accept 33 N) [1] Ecf on second force to be calculated if wrong only because of mistake in first.		4		4	4	

(ii)	Statement (in words): **Either**: least mass if hung from end of bar **Or**: Wire further from mass will be slack (or equiv) when rod about to tilt [1] (If hung from left): $mg \times 0.2$ [m] $= 11.8$ [N] $\times 1.3$ [m] [1] $m = 7.82$ kg [1]				3				1
								3	1
Total				3	6	3	12	8	

Rhodri's answers

(a) The sum of the upward forces on the object is equal to the sum of the downward forces. (not enough)

The sum of the clockwise moments equals the sum of the anticlockwise moments. ✓ X

MARKER NOTE

We need force components in any direction to sum to zero. First mark not gained.

Third mark not gained as it's important to state that the clockwise and anticlockwise moments are taken about the same point.

1 mark

(b) (i) (I)

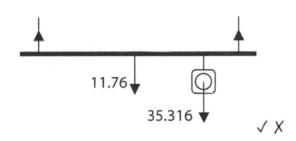

11.76

35.316

✓ X

MARKER NOTE

The second mark is lost for the omission of the unit. Rhodri has given too many significant figures, but this is more likely to be penalised in a question based on experimental results. Best avoided though!

1 mark

(II) Moments about B

$11.8 \times 1.3 + 35.3 \times 0.6 = T_A$ X

$\therefore T_A = 36.5$ N X

$T_A + T_B = 11.8 + 35.3$ ✓ecf

$\therefore T_A + T_B = 47.1$

$\therefore T_B = 47.1 - 36.5 = 10.6$ N ✓ecf

MARKER NOTE

In the first equation Rhodri has written T_A instead of 2.6 [m] $\times T_A$. Writing a force instead of a moment is an error of principle (and a surprisingly common one), so the first and third marks are lost. However, $T_A + T_B = 11.8 + 35.3$ is the correct force equation and is used correctly to find T_B, with ecf on T_A. So second and fourth marks are gained.

2 marks

(ii) Tilting most likely if mass hung from end ✓

$mg \times 0.2 = 11.8 \times 1.3 + 10.6 \times 2.6$ X

mass $= 22$ kg X

MARKER NOTE

Only the first mark awarded for the realisation of where the mass needed to be hung. Rhodri did not realise that the tension in the far wire would be zero and (in desperation?) used the previous value.

1 mark

TOTAL **5 marks / 12**

Ffion's answers

(a) The sum of the forces in any one direction on the object is zero. ✓bod

The sum of the clockwise moments about the pivot is equal to the sum of anticlockwise moments ✓ about the pivot.   ✗

> **MARKER NOTE**
> Ffion should have written 'the sum of the <u>components</u> of the forces in any direction is zero'. This is a subtle point and the examiner awarded the first mark bod.
>
> The third mark was not awarded because there does not have to be a pivot — moments can be taken about <u>any</u> point.
>
> **2 marks**

(b) (i) (I)

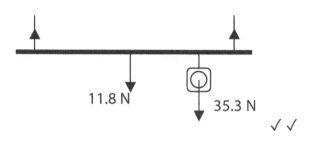

11.8 N   35.3 N

✓ ✓

> **MARKER NOTE**
> A perfect answer.
>
> **2 marks**

(II) $1.3T_A + 24.71 = 1.3T_B$

$T_A + T_B = 47.1\,N$ ✓

$\therefore 1.3T_A + 24.71 = 1.3(47.1 - T_A)$

$\therefore 2.6T_A = 46.52$

$\therefore T_A = 17.9\,N$ and $T_B = 29.2\,N$ ✓✗✓

> **MARKER NOTE**
> The first equation resulted from applying the PoM about the centre of the rod. Since the candidate did not tell us this, any slip in calculating the 24.7(1) N m moment of the spotlight would have made the method almost impossible to follow and most of the marks would have been lost — a high-risk tactic! Note also that it would have been neater to take moments about A or B because the 'perpendicular distance' of T_A or T_B would be zero so only one of the unknown tensions would appear in the equation. As it is, the candidate has solved a pair of simultaneous equations, making only an arithmetical mistake (46.52 instead of 36.52), and losing the third mark.
>
> **3 marks**

(ii) $0.2\,W = 1.3 \times 11.8$ ✗✓

$\therefore W = 76.7\,N$

$mass = \dfrac{76.7}{9.81} = 7.8\,kg$ ✓

> **MARKER NOTE**
> The first mark is lost because the candidate did not make any attempt to explain her reasoning, as required in the question. The analysis was spot on.
>
> **2 marks**

> **TOTAL**   **9 marks / 12**

Section 2: Kinematics

Topic summary

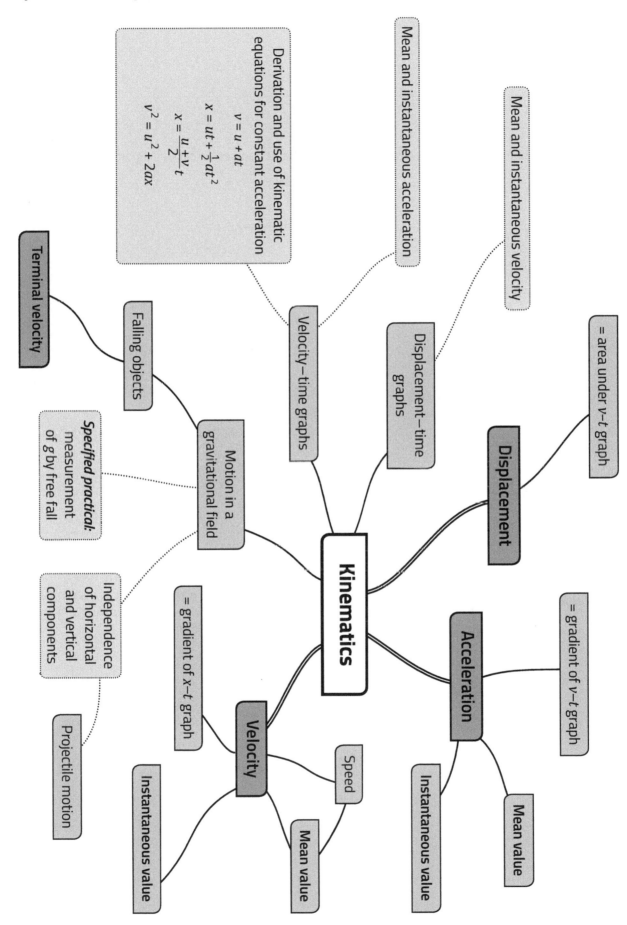

Derivation and use of kinematic equations for constant acceleration

$$v = u + at$$
$$x = ut + \frac{1}{2}at^2$$
$$x = \frac{u+v}{2}t$$
$$v^2 = u^2 + 2ax$$

Mean and instantaneous acceleration

Mean and instantaneous velocity

Terminal velocity

Falling objects

Velocity–time graphs

Displacement–time graphs

= area under v–t graph

Specified practical: measurement of g by free fall

Motion in a gravitational field

Displacement

Independence of horizontal and vertical components

= gradient of x–t graph

Kinematics

Acceleration

= gradient of v–t graph

Projectile motion

Velocity

Speed

Instantaneous value

Instantaneous value

Mean value

Mean value

Q1 (a) Define:

(i) Mean speed. [1]

....

....

(ii) Mean velocity. [1]

....

....

(b) Rhiannon runs from A to C around two sides of a square field (see diagram). She takes 27 s. Calculate:

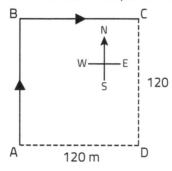

(i) Her mean speed. [1]

....

....

(ii) Her mean velocity. [2]

....

....

Q2 A ball moving to the east at 19 m s^{-1} collides with a vertical wall and bounces back in the opposite direction at 11 m s^{-1}. The time of contact with the wall is 25 ms. Determine the ball's mean acceleration during the collision, and comment on its magnitude. [4]

....

....

....

....

....

....

....

Q3 (a) Derive the following equations for uniformly accelerated motion. If you choose to do so, you may use the sketch graph, adding your own labels.

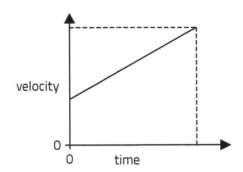

(i) $v = u + at$ [1]

....

....

....

(ii) $x = \dfrac{u + v}{2}t$ [2]

....

....

....

(iii) $x = ut + \tfrac{1}{2}at^2$ [2]

....

....

....

....

(b) Use the equations $v = u + at$ and $x = \dfrac{u + v}{2}t$

to derive the following equation for uniformly accelerated motion.

$$v^2 = u^2 + 2ax.$$ [3]

....

....

....

....

Component 1 Practice questions

Q4 A stone is thrown with a velocity of 15.5 m s^{-1} upwards.

(a) Neglecting forces other than the Earth's gravitational pull, determine

(i) its maximum height above its point of launch, [2]

(ii) the time it takes to reach its maximum height. [2]

(b) Bryn says that the time taken for the ball to reach half its maximum height should be half the time to reach its maximum height. Without further calculation, evaluate this claim. [2]

(c) (i) Calculate the time it takes **from launch** for the stone to be at half its maximum height, on the way back down. [3]

(ii) Calculate its speed at this time. [2]

Q5 The Ilyushin Il-76 is an aeroplane that can be used as a water bomber. It carries a large quantity of water for dropping on to burning areas, e.g. forest fires. A bomber is flying at a height of 100 m at a speed of 120 m s^{-1}. It releases its payload (the water) before it is over the fire.

(a) Explain in terms of the motion of the water why it has to do this. [2]

(b) Calculate how far before the burning area the plane should release the water. [Ignore the effect of air resistance.] [3]

Q6 Huw throws a ball at a speed, u, of 20.0 m s^{-1}, at an angle, θ, of 37° above the horizontal.

(a) (i) Calculate its initial horizontal velocity component. [1]

....

....

(ii) Calculate its initial vertical velocity component. [2]

....

....

(iii) Huw is worried that his answers to (a) (i) and (ii) add up to more than the original speed. Discuss whether or not he ought to be worried. [2]

....

....

....

(b) Determine:

(i) The maximum height gained by the ball. [2]

....

....

(II) The ball's horizontal range (from its point of launch to its return to the same level). [3]

....

....

....

....

(c) Huw reads in an A level maths textbook that the range, R, of a projectile is given by:

$$R = \frac{u^2 \sin 2\theta}{g}$$

(i) Show that the equation is homogeneous (i.e. it works in terms of units). [2]

....

....

....

(ii) Compare the result of using this equation with your answer to (b)(ii). [1]

....

....

Q7 A velocity–time graph is given for a cyclist on a straight road.

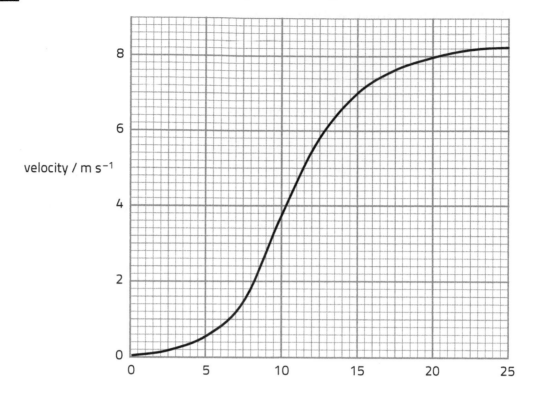

(a) Adding to the graph if you wish to do so, determine:

(i) The magnitude of the cyclist's mean acceleration between 10.0 s and 20.0 s. [2]

....

....

....

(ii) The magnitude of her acceleration at 15.0 s. [4]

....

....

....

....

....

(b) Iolo claims that (a) (ii) could be answered by determining the mean acceleration over the time interval 14.5 s to 15.5 s. Evaluate his claim. [2]

....

....

....

Q8 An idealised velocity–time graph for an electric car travelling between two sets of traffic lights is as follows:

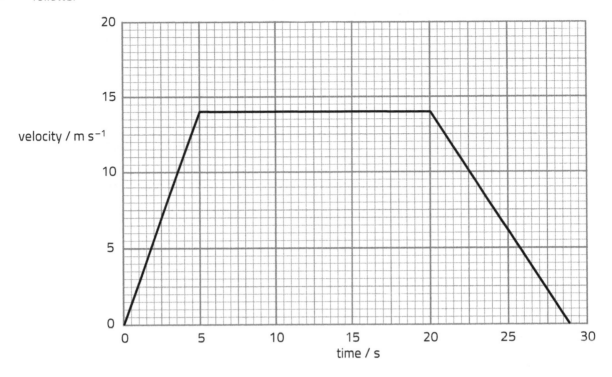

(a) Determine the mean velocity of the car during the motion. [3]

....

....

....

....

....

....

(b) Using the same grid, draw a displacement–time graph for the motion. You will need to add a displacement scale to the grid. [3]

Space for additional calculations, if required.

(c) Sketch an acceleration–time graph for the motion in the space below. Label significant values. [3]

Q9 Helen is sitting in the open carriage of a narrow-gauge railway which is travelling at a constant speed of 8.0 m s^{-1}. She throws a ball vertically upwards (from her point of view) at a speed of 10.0 m s^{-1} (i.e. the actual vertical component of the ball's velocity is 10.0 m s^{-1}) and then catches it.

You should ignore the effects of air resistance for parts (a) – (c) of this question.

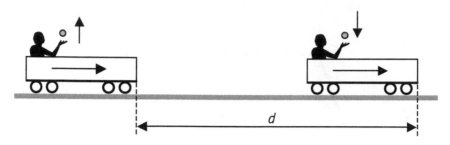

(a) Calculate the distance that the train moves during the time the ball is in the air. [3]

....

....

....

....

....

....

(b) Add to the diagram above to sketch the path taken by the ball during its motion through the air, as seen by an observer standing outside the train. [1]

(c) Explain in terms of the vertical and horizontal motions of the ball why Helen is able to catch the ball in spite of having moved the distance you calculated in part (a). [2]

....

....

....

(d) Discuss how air resistance would have affected the motion of the ball. Think about it from the points of view of Helen and the observer. [3]

....

....

....

....

....

....

Question and mock answer analysis

Q&A 1 A velocity–time graph (with scales removed from the axes) is given for a body moving along a vertical line. Describe what the graph tells us about the body's motion during each stage, AB, BC and CD, in terms of its velocity, its acceleration and its displacement from its starting point (at A).

[6 QER]

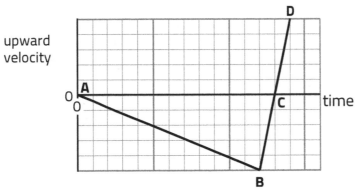

What is being asked

On the face of it, this is a very straightforward question about the interpretation of a velocity–time graph. The thing to ensure, in this QER question, is that all the points are covered, i.e. the sections AB, BC and CD as well as the three aspects of the motion, and all in a well-linked narrative. For a mark in the top band (5–6 marks), you need give a comprehensive description of velocity, acceleration and displacement in all three stages. The middle (3–4 marks) and bottom (1–2 marks) bands have less good coverage of aspects of motion or of the stages, or both.

Mark scheme

| Question part | Description | AOs | | | Total | Skills | |
		1	2	3		M	P
	Indicative content **AB** – velocity: increases from zero; downwards 　　　　acceleration: constant; downwards 　　　　displacement: increases downwards; at an 　　　　　increasing rate **BC** – velocity: downwards; falls to zero 　　　　acceleration: constant; upwards; large compared to 　　　　　over AB 　　　　displacement: more downward displacement; at a 　　　　　decreasing rate; only a little more **CD** – velocity: upwards; increases from zero; ends at the 　　　　　reverse velocity as B 　　　　acceleration: same as over BC 　　　　displacement: decrease in downward displacement; 　　　　　but only small / equal to gain over BC.	2	4		6		
Total		2	4		6		

Rhodri's answers

In AB the body moves downwards, getting faster, therefore it accelerates. Its displacement (down) steadily increases.

In BC the body loses all its velocity. It has a fast acceleration up. Its displacement is still increasing.

In CD the body gets faster again, but going up, with the same acceleration as before. The downward displacement decreases because the body is going up.

MARKER NOTE
There are no glaring mistakes in this answer, but there is a general lack of detail, which makes this a middle-band answer. In AB we ought to have been told that the acceleration is downwards. 'Steadily' is not a good word to describe how the displacement is increasing.
In BC one minor criticism is that accelerations aren't *fast* or *slow*, but *large* or *small*, *high* or *low*. It was not mentioned that the extra displacement is small compared with that over AB.
In CD 'the same acceleration as before' is ambiguous. We also needed something to be said about how much or how little the downward displacement decreases.

4 marks

[**Comment:** Marking QER questions involves judgement. A different examiner might have awarded 3 marks.]

Ffion's answers

Over AB the body's velocity decreases, but actually the body is getting faster. This means that it has an acceleration that is negative and constant (constant gradient). Its displacement is negative and increasing in amount.

Over BC the body is slowing down, though the slope of the graph is large and positive so the acceleration is positive, constant and larger than the acceleration over AB. The negative displacement is still increasing in amount but not by very much as the area under the graph is small here.

Over CD the body's velocity is now increasing. The acceleration is the same as in BC. The body's negative displacement decreases a little.

MARKER NOTE
This candidate's answer would have been excellent if she had used *upwards* instead of *positive* and *downwards* instead of *negative*. The rather confused first sentence suggests that some conceptual problem about vector quantities. It would have been clearer if she had said that the magnitude of the velocity increases in AB.
One small omission is that no comment has been made about rates at which displacement changes in any stage.
The question required descriptive statements rather than explanations, but the candidate has justified her answer (for BC) by mentioning graph gradient and area under graph. These two short remarks do tend to enhance her answer, but too much explanatory material would have clogged up the description asked for.
The good coverage of a, v and x in all three regions makes this a top-band answer.

5 marks

Q&A 2

(a) In an experiment to determine a value for g, a student drops six marbles, one at a time, from an eighth-floor window of a tall building. Another student uses a stopwatch to time how long they take to reach the ground (in a roped-off area). The height dropped is measured as 24.30 m ± 0.01 m. The times recorded are:

2.31 s   2.47 s   2.26 s   2.42 s   2.35 s   2.51 s

Calculate a value for g from these results, along with its absolute uncertainty, giving your working.

[6]

(b) Comment on the value for g and its uncertainty found in part (a), and briefly discuss likely causes of error and uncertainty.

[4]

What is being asked

This question directly tackles experimental skills. It is set around one of the set practicals: the measurement of g by free fall. Part (a) involves a standard use of experimental data to determine a result with its associated uncertainty. The uncertainty in one variable (the drop height) is given; the uncertainty in time must be calculated from the raw results. Part (b) requires an evaluation of the final result.

Mark scheme

Question part	Description	AOs 1	AOs 2	AOs 3	Total	Skills M	Skills P
(a)	mean time = 2.39 s [1] $$g = \frac{2h}{t^2} \text{ (or by implication) [1]}$$ $g = 8.51$ m s^{-2} or 8.5 m s^{-2} [1] $$\Delta t = \frac{2.51 - 2.26}{2} \text{ (or by impl.) [1] } [= \pm 0.13 \text{ s}]$$ $$p_g = 2 \times \frac{0.13 \text{ (ecf)}}{2.39} \times 100 \text{ or by impl. [1] } [= 11\%]$$ $\Delta g = \pm 0.9$ m s^{-2} **or** 0.94 m s^{-2} if g to 2 d.p. [1] ***Alternative for last 3 marks*** Either max or min g calculated [9.52 m s^{-2}; 7.71 m s^{-2}] ✓ g uncertainty = (max − min) / 2 , or equiv. ✓ $\Delta g = \pm 0.9$ m s^{-2} **or** 0.91 m s^{-2} if g to 2 d.p. ✓	6				4	6
(b)	The value for g is low or the [%] uncertainty is large [1] The standard value of g isn't allowed by the uncertainty [1] Large [random] uncertainty expected when timing such short intervals by eye with stop-watch. [1] Low value of g probably due to a [systematic] error, e.g. air resistance [over long fall] **or** in timings. [1]		4			4	4
Total		6	4		10	8	10

Rhodri's answers

(a) $2.31 + 2.47 + 2.26 + 2.42 + 2.35 + 2.51 = 14.32$

$14.32 \div 6 = 2.39$ ✓

$24.3 = \frac{1}{2} \times g \times 2.39^2$

$g = \frac{2 \times 24.3}{2.39^2}$ ✓ $= 8.5 \text{ m s}^{-2}$ ✓

$2.51 - 2.31 = 0.2$ ✗

$0.2 \div 2 = 0.1$, $\frac{0.1 \times 100}{2.39} = 4.18\%$

$4.18\% \times 2 = 8.36\%$ ✓ ecf

$\frac{9.51 - 8.51}{2} = \pm 0.71 \text{ m s}^{-2} \text{ in } g$ ✗

> **MARKER NOTE**
> Rhodri has followed a correct procedure, showing all his steps. It would have been safer, though, to give a few more *words* stating what his figures represent. He has made a slip in calculating his uncertainty in the times, choosing 2.31 s instead of 2.26 s as his lowest; fourth mark lost. Since he has given g to only 1 dp, it is wrong to give its uncertainty to 2 dp; sixth mark lost.
> **4 marks**

(b) The value of g is too low, since the right value is 9.8 m s^{-2}. ✓ This may be because of air resistance (drag). ✓ A stopwatch isn't good enough for such short time intervals. (not enough)

> **MARKER NOTE**
> Rhodri's first sentence gains the first (easy) mark, and his second sentence, the fourth mark. He's missed the second mark, not commenting that his range of allowed values doesn't include 9.81 m s⁻². He realised that the stop-watch technique isn't really suitable, but hasn't linked it with uncertainties, so misses the third mark.
> **2 marks**

Total **6 marks / 10**

Ffion's answers

(a) Mean of times $= \dfrac{2.31 + 2.47 + 2.26 + 2.42 + 2.36 + 2.51}{6}$

$= 2.39 \text{ s}$ ✓

$x = ut + \frac{1}{2}at^2$

$u = 0 \rightarrow a = \dfrac{2x}{t^2} = \dfrac{2 \times 24.3}{2.39^2}$ ✓ $= 8.5 \text{ m s}^{-2}$ ✓

Highest possible acceleration is for 2.26 s

so max $a = \dfrac{2x}{t^2} = \dfrac{2 \times 24.3}{2.26^2} = 9.51 \text{ m s}^{-2}$ ✓

Uncertainty $= \dfrac{9.51 - 8.51}{2}$ ✗ $= 0.5 \text{ m s}^{-2}$ ✗

(no ecf) ✗

> **MARKER NOTE**
> This candidate's answer is very clearly and competently set out. She has chosen the easy way to calculate the absolute uncertainty in g, but has made a serious error in the last step: the uncertainty is 9.51 − 8.51, not
> $\dfrac{9.51 - 8.51}{2}$.
> Presumably she was getting confused with methods that do require division by 2. The last mark cannot be awarded ecf because the mistake is one of principle, and not just a slip. The last two marks are lost.
> **4 marks**

(b) Even the highest possible acceleration is less than the accepted value of 9.81 m s^{-2}. ✓✓. This might be because the timekeeper stops the stop-clock too late every time, ✓ or it might be because air resistance really does reduce the acceleration. In any case, a 0.1 s error in timing such a small interval makes for a large percentage uncertainty. ✓
A stopwatch isn't good enough for such short time intervals.

> **MARKER NOTE**
> Although she has used different words from those of the mark scheme, Ffion's first sentence gains both of the first two marks. Her second sentence contains more than enough to give her the fourth mark. Her next sentence gains her the third mark. A competent answer.
> **4 marks**

Total **8 marks / 10**

Section 3: Dynamics

Topic summary

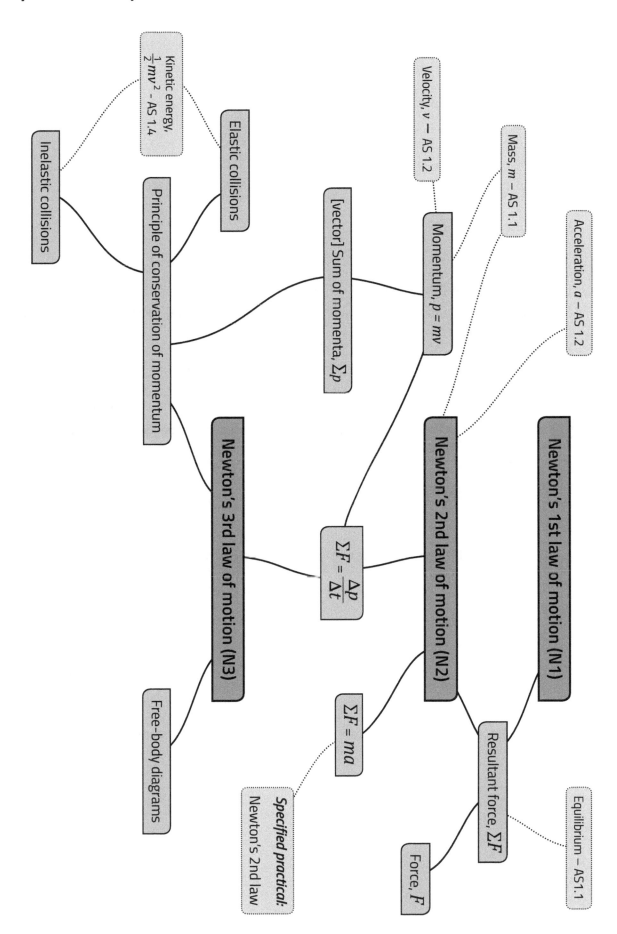

Q1   Newton's third law of motion is about pairs of forces. In part (b) of this question they will be called 'Newton's third law partner forces'.

(a)   State Newton's third law of motion.   [1]

....

....

....

(b)   The diagram shows a book at rest on a table.

book →

(i)   **Show on the diagram** the two main forces on the book, using labelled arrows.   [2]

(ii)   For each of the two forces you have labelled, state the body on which the Newton's third law partner force acts.   [2]

....

....

Q2   A teacher asks two students to write Newton's second law of motion as an equation. Bethan, a GCSE student, writes:

$$F = ma.$$

Angharad, an A level student, writes:

$$F = \frac{\Delta p}{t}.$$

(a)   Defining all the symbols, explain how these two equations are consistent with each other.   [3]

....

....

....

....

(b)   The force that a gas exerts on the walls of its container arises because the gas molecules hit the walls and bounce off. Explain which of the two equations is more useful in this context.   [2]

....

....

....

....

Q3 A box of mass 28 kg is dragged along level ground in an easterly direction by means of a rope.

The box has a constant acceleration. As it slides a distance of 2.7 m its velocity increases from 1.5 m s^{-1} to 2.1 m s^{-1}.

rope

(a) Calculate the resultant force on the block. [3]

....

....

....

....

....

(b) During the acceleration the rope exerts a force of 18.2 N on the box. Determine the magnitude and direction of the frictional force that **the box exerts on the ground**, giving your reasoning. [3]

....

....

....

....

....

Q4 Two trolleys, A and B, each of mass 12 kg, are linked by a chain of mass 2.0 kg. They are pulled by a rope and their acceleration is 0.75 m s^{-2}. The frictional force on each trolley is 5.0 N.

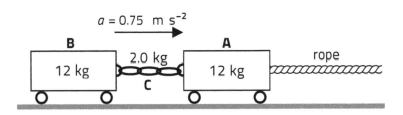

Explaining your reasoning, calculate:

(a) The force exerted by the rope on trolley **A**. [2]

....

....

....

(b) The force exerted by trolley **B** on the chain, **C**. [3]

....

....

....

....

Q5 A body of mass 4.0 kg is acted upon by forces in the horizontal plane, as shown.

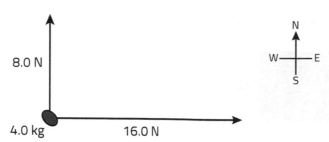

Determine the magnitude and direction (as a bearing) of the body's horizontal acceleration. You may add to the diagram or draw a separate vector diagram in the space to the right of it. [4]

....

....

....

....

....

....

Q6 The diagram shows a steel ball attached by stretched springs to fixed anchorages. The ball is held in position by an electromagnet (not shown). The tension in each spring is 6.0 N.

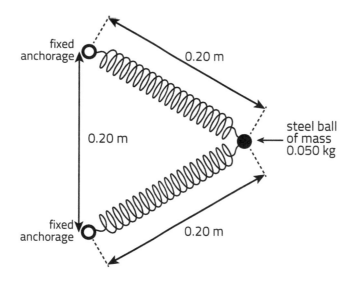

(a) Calculate the ball's acceleration just after the electromagnet is turned off. [4]

....

....

....

....

....

....

(b) Give two reasons why the ball's acceleration will decrease as the ball moves. [Resistive forces are negligible.] [3]

....

....

....

....

Q7 An empty supermarket trolley is pushed and released so that it nests with two identical trolleys that are already nested and moving more slowly in the opposite direction (see diagram). After the collision, the three trolleys move off together with the same velocity.

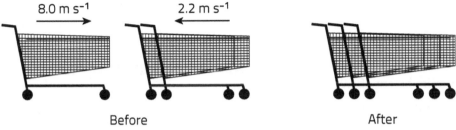

8.0 m s^{-1} 2.2 m s^{-1}

Before After

(a) Determine the velocity of the trolleys after the collision. [3]

....

....

....

....

....

(b) Express the total final kinetic energy of the three trolleys as a fraction of their total initial kinetic energy. [3]

....

....

....

....

(c) A student claims that the 'lost' kinetic energy has been given to the atoms of the trolleys. Discuss this claim. [2]

....

....

....

Q8 A dumbbell consists of two metal weights, each of mass 2.5 kg, separated by a light rod. The dumbbell is rotating about its centre, so that the speed of the weights is 5.0 m s⁻¹.

Calculate the total kinetic energy and the total momentum of the dumbbell. Explain your reasoning. [4]

Q9 A ball of mass 0.220 kg is dropped on to a floor from a height of 2.00 m. It bounces back up, reaching a maximum height of 1.20 m.

(a) Show that the ball's change in momentum during the bounce is roughly 2.5 N s. [5]

(b) The ball is in contact with the ground for 150 ms. Calculate the mean resultant force on the ball during the bounce. [2]

(c) Explain why the mean force exerted by the ground on the ball during the bounce must be roughly 2 N greater than the mean resultant force on the ball calculated in (b). [2]

Q10 Two gliders of unequal mass approach each other on a level air-track, as shown.

After the collision the 0.25 kg glider is moving with a velocity of 0.107 m s^{-1} to the right.

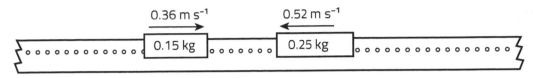

(a) Determine the velocity of the 0.15 kg glider after the collision. [3]

....

....

....

....

....

(b) Show that the collision is inelastic. [3]

....

....

....

....

....

....

Q11 A molecule of mass 6.6×10^{-27} kg travelling at 2 500 m s^{-1} bounces round the inside of an otherwise empty box as shown. The collisions with the walls are elastic and direction is always at 30° to side **XY**.

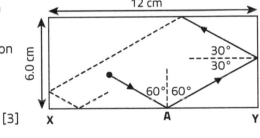

(a) Calculate the momentum change of the molecule at **A**. [3]

....

....

....

....

(b) Calculate the mean force the molecule exerts on the side XY. [Hint: Consider the time taken before the molecule hits the side **XY** again.] [3]

....

....

....

....

Question and mock answer analysis

Q&A 1 A cyclist, Carys, freewheels (rides without pedalling) down a hill. Carys and her bike (C-and-B) have a total mass of 78 kg, and are shown on the diagram as a single rectangle.

(a) **Add labelled arrows to the diagram** to show the forces acting on C-and-B.   [3]

(b) Calculate the component of the gravitational force that accelerates C-and-B.   [2]

(c) Carys's speed increases from 4.5 m s^{-1} to 11.3 m s^{-1} over a time of 12.0 s. Calculate the mean **resistive** force on C-and-B, giving your reasoning.   [3]

(d) Carys believes that the acceleration must be greater towards the end of the 12.0 s interval than at the beginning. Evaluate this belief.   [3]

What is being asked

The motion, or lack of it, of objects on inclined planes is a common setting for AS and A level questions. Knowledge from Sections 1.1 and 1.2 as well as 1.3 is needed here. In part (a), you are expected to remember that gravity (or weight) acts vertically downwards on an object, a flat surface always exerts a force at right angles on an object in contact and that friction will act in a manner which opposes motion. This is classed as AO1 even though the knowledge must be applied to this situation, as you will certainly have encountered the set-up before. Taking components in (b) is a skill from Section 1.1, and calculating acceleration, which is required in (c), is from 1.2, this time applied to forces and motion, a 1.3 concept. (b) and (c) are AO2 questions. Examiners very often pose a supposed student's statement and ask for comments; this is a standard way of setting an AO3 question. In this type of question, there is no mark for just saying that Carys is wrong (or right) but this is required as part of the answer.

Mark scheme

Question part			Description	AOs			Total	Skills	
				1	2	3		M	P
(a)			Downward arrow labelled *weight* (or equiv) [1] Arrow normally 'up' from surface labelled *normal contact force* (accept *normal reaction*) [1] Arrow up slope labelled *resistive force* or equiv [1] No penalties for oddly *positioned* arrows. [–1 mark lost per additional incorrect force]	3			3		
(b)			Multiplication by sin 6.0° [1] [78 × 9.81 × sin 6.0° =] 80 N [1]		2		2	2	
(c)			Mean acc = $\dfrac{11.3 - 4.5}{12.0}$ [m s⁻²][1][= 0.567 m s⁻²] Mean resultant force = 78 × 0.567 N [1] [= 44.2 N] Mean resistive force = [80 N – 44 N] = 36 N [1] Full ecf on 80 N.		3		3	1 1	
(d)			Resistive force [**or** air resistance] increases as speed increases [1] So resultant force decreases [1] So acceleration decreases **and** Carys wrong. [1]			3	3		
Total				3	5	3	11	4	

Rhodri's answers

(a)

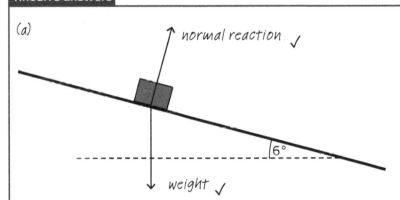

normal reaction ✓

6°

weight ✓

MARKER NOTE
Rhodri gains marks for correctly marking on the weight and 'normal reaction'. He omits the resistive force and so doesn't gain the third mark.
Rhodri has made no attempt to put the tails of the arrows at the points where the forces act, but it is not easy to do so in this case, and there was no penalty for not doing so.
2 marks

(b) Weight = 78 × 9.81 = 765.18 N
Comp down slope = 765.18 × cos 6° X
= 761 N X no ecf

MARKER NOTE
Rhodri has multiplied the weight by cos 6° rather than by sin 6° (or cos 84°). This is a serious error and there is no separate mark for calculating *mg*.
0 marks

(c) Change in velocity = 11.3 – 4.5
= 6.8 m s⁻¹
so accel $\dfrac{6.8}{12}$ = 0.56666 m s⁻² ✓
so force = 78 × 0.56666 = 44.2 N ✓ bod

MARKER NOTE
Rhodri has not calculated the resistive force. He correctly calculated the resultant force, but has not stated that it *is* the resultant force. Examiner gave benefit of doubt here and so he obtains the first two marks.
2 marks

(d) Carys is obviously right. She gets faster and faster as she goes down the slope. X

MARKER NOTE
Rhodri appears to be confusing acceleration with speed or velocity. No credit given.
0 marks

Total	4 marks / 11

Ffion's answers

(a)

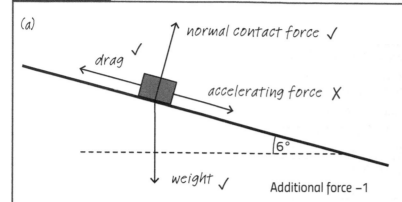

normal contact force ✓

drag ✓

accelerating force ✗

6°

weight ✓

Additional force −1

MARKER NOTE
Ffion gains marks for correctly marking on the weight and the normal contact force (a preferable term to 'normal reaction'). She also marks the weight but also includes an 'accelerating force' which is not a separate force but is either a component of the weight or the resultant of this component and the drag. This arrow shouldn't be there and she is penalised 1 mark.

2 marks

(b) Weight = 78 x 9.81

accelerating force $= 78 \times 9.81 \times \sin 6°$ ✓

$= 80\,N$ ✓

MARKER NOTE
Clear and correct. Both marks obtained.

2 marks

(c) Acc'n due to 80 N $= \dfrac{80}{78} = 1.03\ m\ s^{-2}$?

actual acc'n $= \dfrac{6.8}{12} = 0.57\ m\ s^{-2}$ ✓

Therefore dec'n due to resistive force = 1.03 − 0.57 = 0.46 m s⁻² ?

Thus resistive force $= 78 \times 0.46 = 36\,N$ ✓

MARKER NOTE
It is no co-incidence that Ffion's procedure gives the right answer. Yet only one of her three accelerations (the 'actual acceleration') has any physical meaning. The reasoning should have been conducted in terms of forces. 1 mark withheld.

2 marks

(d) She is going fastest at the end, so air resistance is greatest then ✓ so Carys is wrong. ✗

MARKER NOTE
The 3 marks allotted suggest that a full answer is needed. Ffion's, though it starts well, lacks the middle step of considering the resultant force. There is no mark for just saying that Carys is wrong.

1 mark

| Total | 7 marks / 11 |

Section 4: Energy concepts

Topic summary

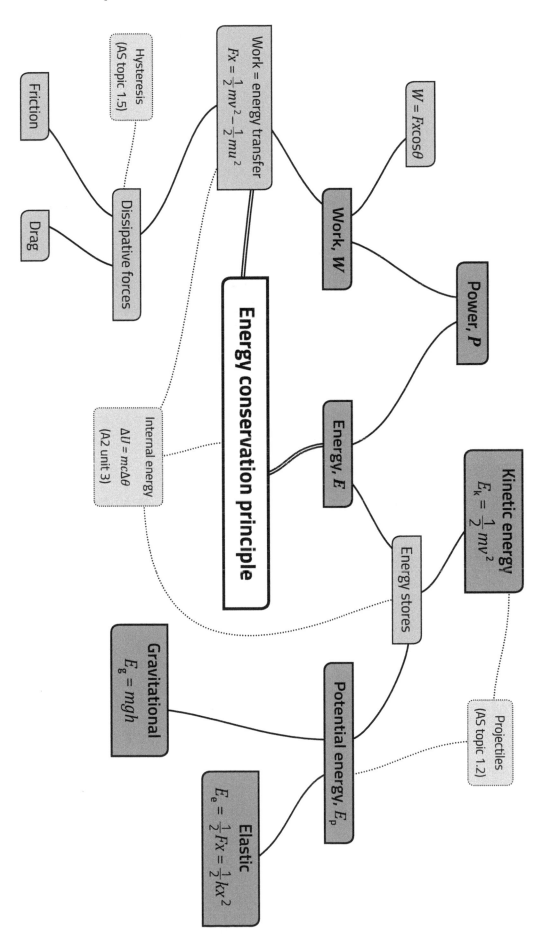

Hysteresis
(AS topic 1.5)

Friction

Drag

Work = energy transfer
$Fx = \frac{1}{2}mv^2 - \frac{1}{2}mu^2$

$W = Fx\cos\theta$

Dissipative forces

Work, W

Power, P

Internal energy
$\Delta U = mc\Delta\theta$
(A2 unit 3)

Energy conservation principle

Energy, E

Kinetic energy
$E_k = \frac{1}{2}mv^2$

Energy stores

Gravitational
$E_g = mgh$

Potential energy, E_p

Projectiles
(AS topic 1.2)

Elastic
$E_e = \frac{1}{2}Fx = \frac{1}{2}kx^2$

Q1 The kilowatt-hour (kW h) is a unit of energy.

(a) Use the definition of power to explain why the kW h is a unit of energy. [1]

....

....

(b) The battery of an electric car is rated as 96 kW h. Express this in joule (J) in standard form. [2]

....

....

....

Q2 (a) State the principle of conservation of energy. [1]

....

....

(b) A stone of mass 0.150 kg is thrown diagonally upwards with a speed of 50.0 m s^{-1}.

(i) Assuming that air resistance is negligible, use the principle of conservation of energy to calculate the stone's speed at its maximum height of 31.9 m. [3]

....

....

....

....

....

(ii) Without calculation, explain whether the answer to (i) would be different if the mass of the stone were different. [2]

....

....

....

Q3 A cyclist freewheels down a hill which has a gradient of 5.0°. After travelling 200 m from rest, her speed is 12.0 m s^{-1}. The combined mass of the cyclist and bicycle is 85 kg.

(a) Calculate the loss in gravitational potential energy. [2]

....

....

....

(b) Using an energy argument, calculate the mean resistive force acting. [2]

....

....

....

Q4 The Highway Code gives the typical braking distance for a car travelling at a speed of 60 mph (26.7 m s^{-1}) on a dry road with good tyres as 55 m.

(a) Calculate the braking force assumed by the Highway Code for a car of mass 1200 kg. [3]

....

....

....

....

....

(b) The braking distance for cars travelling at 30 mph, 50 mph and 70 mph are given as 14 m, 38 m and 75 m respectively. Evaluate whether the Highway Code assumes a constant braking force. [3]

....

....

....

....

....

(c) The typical braking distance figures given in the Highway Code are the same for all cars. John states that this means that all cars are assumed to have the same braking force. Discuss whether John's statement is correct. [2]

....

....

....

Q5 The Eduqas Terms and Definitions booklet defines power as 'the work done per second or the energy transferred per second'.

(a) Use a definition of work or energy to explain why the two alternative parts of the definition are equivalent. [2]

....

....

....

(b) John states that in some situations either alternative is useful but sometimes only one of them is appropriate.

The Sun's power output is approximately 3.8×10^{24} W; a horse dragging a log has a useful power output of about 600 W. Use these examples to illustrate John's statement. [3]

....

....

....

....

Q6 A vertical spring with a spring constant 32.0 N m^{-1} is clamped at its top end. A load of mass 0.600 kg is attached to the bottom of the spring, at a height of 0.400 m above the laboratory bench and supported so that the spring is not stretched. The load is then released.

(a) Show that the initial gravitational potential energy (from bench height) is approximately 2.4 J. [2]

(b) Calculate the loss in gravitational potential energy, the elastic potential energy and the kinetic energy when the load has fallen 0.184 m. [3]

(c) Use your answer to (b) to calculate the speed of the load when it has fallen 0.184 m. [2]

(d) Show that the lowest point that the load reaches is 0.032 m above the bench. [3]

(e) A student correctly works out that the speed, v, of the load when it has fallen by a distance h can be worked out using the equation:

$$mgh = \tfrac{1}{2}kh^2 + \tfrac{1}{2}mv^2$$

(i) Explain this equation in terms of energy transfer. [2]

(ii) Use the equation above to calculate the values of h at which $v = 1.00$ m s^{-1}. [3]

Q7 (a) The work, W, done by a force of magnitude, F, which moves its point of application by a distance d in the same direction as F is given by $W = Fd$.

Use definitions of velocity and power to show that the power, P, transferred by a force F moving at a velocity v is given by $P = Fv$. [2]

....

....

....

....

(b) The aerodynamic drag, D, on a new-model electric SUV is given by $D = kv^2$, where k is a constant, with a value of 0.4 in SI units.

(i) The unit of k can be written N $(m\ s^{-1})^{-2}$. Express this in base SI units in the simplest terms. [2]

....

....

....

(ii) Calculate the power transferred by the SUV's engine when the car is travelling at a steady speed of 30 m s^{-1}. [3]

....

....

....

....

....

(iii) An advert for the SUV says that, with a 100 kW h battery, the range of the car is 900 km. Evaluate this claim. [You should make reasonable assumptions about the efficiency of the engine drive system and the driving speed.] [4]

....

....

....

....

....

....

....

Component 1 Practice questions

Q8 A traveller in the Arctic drags a laden sledge of total mass 210 kg a distance of 7.0 km across a frozen surface, in a straight line at a constant speed of 1.4 m s^{-1}. To do so, he applies a constant force of 83 N to the sledge via a tow-rope, which is inclined at an angle of less than 5° to the horizontal.

(a) Calculate the work done on the sledge by the traveller. [2]

....

....

....

(b) Considering the number of significant figures in the data, explain why, to answer part (a), it was not necessary to know the actual angle of the tow-rope to the horizontal. [2]

....

....

....

(c) State the value of the frictional force between the sledge and the ice. [1]

....

(d) For most of the 7.0 km journey, the sledge gains no kinetic energy. Explain why not and state the nature of the energy transferred to the sledge and ice. [2]

....

....

....

(e) At the end of the journey, the friction between the sledge and the ice causes the sledge to come to a stop.

(i) State the energy change that occurs during this process. [1]

....

(ii) Calculate the distance travelled by the sledge in stopping. [2]

....

....

....

(f) At the start of the journey, the traveller applies a force of 105 N to the sledge to accelerate it from rest to its steady speed. Aled claimed that most of the work done in the process went into the kinetic energy of the sledge. Evaluate this claim. [2]

....

....

....

Q9 Bethan, a hill farmer in mid-Wales, installs a small electrical wind turbine, with 6.0 m blades, to provide power for her farm. On one particular day, the wind speed is 15 m s^{-1}.

(a) She calculates that the mass, m, of air interacting with the blades of the wind turbine per second is given by:

$$m = \pi r^2 v \rho$$

where r is the radius of the blades, v is the wind speed and ρ the density of the air.

(i) Show that this equation is homogeneous, i.e. that the units are the same on the two sides. [2]

(ii) By considering the kinetic energy of the air, show that the power input, P_{IN}, to the wind turbine can be calculated using the formula:

$$P_{IN} = \frac{1}{2}\pi r^2 \rho v^3$$

[2]

(iii) The efficiency of the wind turbine itself is 56%. It is coupled to an electrical generator of efficiency 95% via a gear-box of efficiency 85%. Calculate the electrical power output, in kW, of the generator, to an appropriate number of significant figures.
[ρ_{air} = 1.3 kg m^{-3}]

[4]

(b) Bethan knows that the output from the wind turbine is variable so plans to install rechargeable batteries to store energy for use in non-windy conditions, removing the need to buy electricity from the grid. Her mean power requirement is 75% of the value you calculated in (a)(iii). She knows that the mean wind speed on her farm is 7.5 m s^{-1}. She believes that the mean power output of the generator is therefore half the value that you calculated in (a)(iii), which would not be enough to provide all the power needs of the farm. Evaluate whether she is correct. [3]

Question and mock answer analysis

Q&A 1

(a) The diagram shows a steel sphere resting against a compressed spring. The knob is released, allowing the spring to expand quickly, causing the ball to travel along the track from A, upwards to D and horizontally towards E.

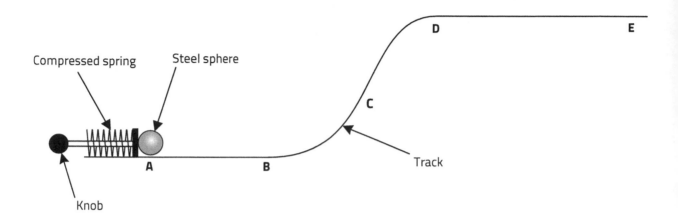

Explain the motion of the sphere in terms of forces and the energy transfers that occur from the moment the spring starts to expand. [6QER]

(b) A zip-wire ride in an adventure park has a length of 200 m and a fall of 25 m.

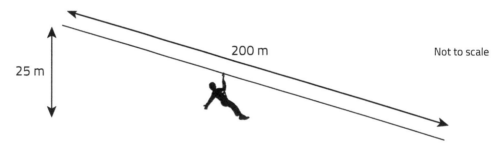

A rider of mass 70 kg achieves a speed of 15 m s⁻¹ just before reaching the bottom of the ride.

Calculate the mean resistive force on the rider during the ride. [4]

What is being asked

This question is about energy transfers.

Part (a) is a *Quality of Extended Response* (QER) question, which is marked in terms of both content and clarity of the answer. It has both AO1 and AO2 aspects. The examiner is looking for an explanation of the forces that do work on the ball, causing named energy transfers. See page 8 for a general description of QER questions. In this case, examiners are looking for both knowledge of the energy stores (gravitational potential, elastic potential, kinetic, etc.) and the mechanism by which the transfer occurs, i.e. work by an identified force. Low-level answers (1–2 marks or 3–4 marks) would miss out some of the energy transfers, not appreciate that work is the means by which the energy transfers occur or incompletely identify the forces involved.

Part (b) is a calculation question following on from the theme of work and energy transfers in part (a), which is also AO1 and AO2, involving basic mathematical reasoning.

Mark scheme

Question part			Description	AOs			Total	Skills	
				1	2	3		M	P
(a)			**Indicative content**						
			• Energy transfers						
			• Identification of energy forms						
			• Elastic → kinetic						
			• Kinetic → gravitational potential						
			• Energy loss as internal energy / heat						
			• Transfers linked to positions						
				2	4		6		
			Forces and energy transfers						
			• Work as the agency of energy transfer						
			• Force from spring: elastic → kinetic						
			• Gravitational force: kinetic → gravitational						
			• Dissipative forces: kinetic → internal / heat						
			• Friction or air resistance as dissipative force						
			• Molecular explanation of friction or air resistance.						
(b)			mgh used to calculate loss in E_p or by impl. [e.g. 17167.5 (J) seen] [1]						
			$\frac{1}{2}mv^2$ used to calculate gain in E_k or by impl. [e.g. 7875 (J) seen] [1]	2	2		4	4	
			Total loss in $E_p + E_k$ calculated, ecf, or by impl. [e.g. 9292.5 (J) seen] [1]						
			Resistive force $= \left[\dfrac{9292.5}{200}\right] = 46(.5)$ N ecf [1]						
Total				**4**	**6**	**0**	**10**	**4**	

Rhodri's answers

(a) At the start, the energy is all potential energy in the spring. When the spring is released, it pushes the ball, giving it kinetic energy. As it moves, the ball has kinetic energy between A and B, then it gains potential energy and loses kinetic energy to D. After D it slows down, losing kinetic energy because of friction, until it stops. Then it just has potential energy.

(b) PE $= 70 \times 9.81 \times 25 = 17167.5$ J ✓

KE $= \frac{1}{2} \times 70 \times 15^2 = 7875$ J ✓

$v^2 = u^2 + 2ax$, $15^2 = 0^2 + 2a \times 200$,

so $a = \dfrac{15^2}{2 \times 200} = 0.5625$ m s^{-2} ✗

$F = ma = 70 \times 0.5625 = 39.375$ N

✗

no ecf

MARKER NOTE

Rhodri identifies most of the energy forms, omitting internal energy / heat, but doesn't distinguish between elastic and gravitational potential energy. He mentions two forces (spring push and friction) but does not mention the role of work in transferring energy. This is a low-level (bottom band) answer. His answer would be improved by identifying work against the force of gravity in the transfer from kinetic to gravitational potential energy and similarly work against friction in the transfer to internal / heat energy.

2 marks

MARKER NOTE

Rhodri starts well. He uses the hint of part (a) well and correctly calculates the change in gravitational potential and kinetic energies and so gains the first two marks. He doesn't explicitly state mgh or $\frac{1}{2}mv^2$ but he has clearly used them, as required in the mark scheme. There is no penalty for significant figures.

The remaining two marks are not given as the work is not relevant. Rhodri has in fact calculated the mean resultant force on the rider. There is an alternative method of answering the question, which involves Rhodri's calculation:

1 Calculate the component of the gravitational force down the wire (assumed straight) using $mg\sin\theta$ (85.8 N).

2 Subtract the resultant force Rhodri has calculated to give the 46.4 N, which is the correct answer.

However Rhodri cannot obtain marks for two partial answers.

2 marks

| **Total** | **4 marks / 10** |

Component 1 Practice questions

Ffion's answers

(a) The forms of energy involved are: elastic potential energy (EPE), gravitational potential energy (GPE), kinetic energy (KE) and heat (H). To start with the spring has EPE because it is compressed. When it expands, it exerts a force on the ball, which does work and the ball gains KE (between A and B). Between B and D, the ball rises so it gains GPE and loses KE. The loss in KE is equal to the gain in GPE. So at D the ball has less KE than at B. As it rolls from D towards E it slows down because it hits air molecules (i.e. air resistance), so it loses more KE and the air gains heat energy (H).

(b) Energy loss = $mgh - \frac{1}{2}mv^2$

$= 70 \times 9.81 \times 25 \checkmark - \frac{1}{2} \times 70 \times 15^2$

$= 1420 \text{ J } (3 \text{ sf}) \quad \text{X} \checkmark \text{ ecf}$

So work against friction $= 1420 = F \times 200$

So $F = \frac{1420}{200} = 7.1 \text{ N} \checkmark \text{ ecf}$

MARKER NOTE

Ffion correctly identifies all the relevant stores of energy. In an A2 question, internal energy would be expected instead of heat but this is accepted here. She also correctly identifies work as the agency of energy transfer and, in two of the transfers, names the force involved. The role of air molecules as applying the force of air resistance is a good point to make.

The answer could be improved in two ways:

- By identifying the role of gravity in the transfer of kinetic energy to gravitational potential energy, i.e. work done against gravity.
- By noting that air resistance (and rolling friction) is significant all through the journey, not just after D.

This is a high-level answer.

5 marks

MARKER NOTE

This is almost a perfect answer, with just one slip, which has cost Ffion a mark.

Ffion sets out the method of calculating the energy loss concisely, involving both the gravitational potential energy (for which she gets the first mark) and the kinetic energy. Unfortunately she has made the slip of not using the $\frac{1}{2}$ in calculating the KE, so she has not used $\frac{1}{2}mv^2$ and hence, she misses the second mark.

However, she obtains the third mark for combining the change in PE and KE on the ecf principle, as the error was just an arithmetic slip. The last mark is also given on the ecf principle although Ffion's answer of 7.1 N is wrong..

3 marks

| Total | 8 marks / 10 |

Section 5: Circular motion

Topic summary

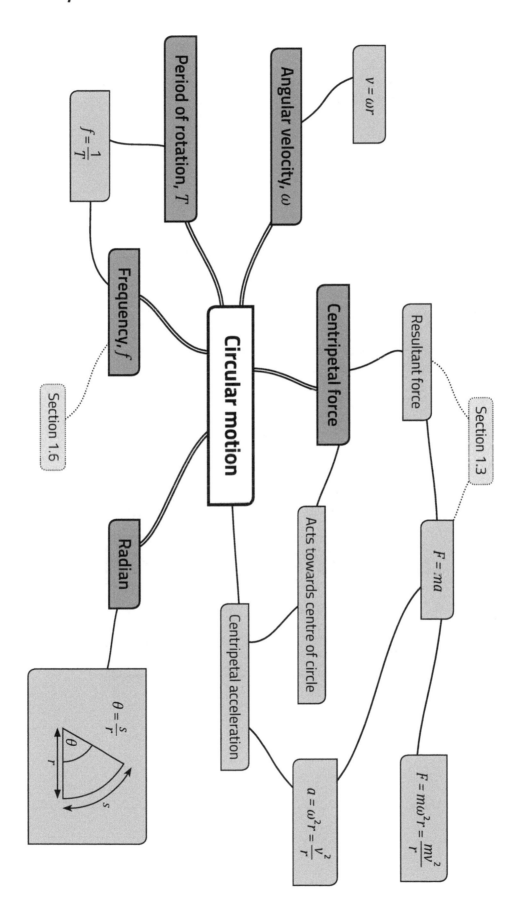

Q1 An angle, θ, has magnitude 1.2 radians. Explain the meaning of this statement, with the aid of a diagram, and explain how an angle in radians can be expressed in degrees (°). [2]

....

....

....

Q2 An object executes circular motion. Define the terms *period of rotation* and *frequency* and explain the relationship between them. [3]

....

....

....

....

Q3 (a) Define *angular velocity* for an object in circular motion. [1]

....

....

(b) The maximum rate of rotation of a washing machine is advertised as 1400 revolutions per minute. Calculate the maximum angular velocity of the washing machine. [2]

....

....

....

Q4 An object of mass 65 kg travels in a circle of radius 4.5 m and at a constant speed, v, of 23.2 m s^{-1} on ice. A light rope connects the object to the centre of the circle.

(a) Calculate the tension in the light rope. [2]

....

....

....

(b) The rope is pulled towards the centre to make it shorter. Explain why this makes the object move more quickly. [2]

....

....

....

....

Q5 A car travels at a constant speed along a flat road and round a curve whose radius of curvature is 24.0 m.

(a) State the direction of the car's acceleration and the resultant force on the car. State what provides this force. [3]

....

....

....

(b) When the magnitude of the centripetal force exceeds the weight of the car, the car will skid and lose control. Calculate the maximum speed that the car can travel around the curve safely. [3]

....

....

....

....

....

(c) Joe claims that when the radius of curvature of the bend increases, the car can travel faster around the curve but that the maximum angular velocity has to decrease. Determine whether, or not, Joe's claims are accurate. [4]

....

....

....

....

....

....

....

Q6 (a) Use the data relating to the planet Saturn to answer the following questions:

Planet	Orbit radius (km)	Mass (kg)	Orbit period (year)	Planet radius (km)	Day length (hour)
Saturn	1.43×10^9	5.68×10^{26}	29.5	60 000	10.7

(i) Calculate Saturn's angular velocity about the Sun. [2]

(ii) Calculate Saturn's angular velocity about its North–South axis. [2]

(b) (i) Calculate Saturn's orbital speed (about the Sun). [2]

(ii) Calculate the rotational speed of a point on Saturn's equator. [2]

(c) (i) Calculate the centripetal acceleration and force acting on the planet Saturn and state what provides this force. [4]

(ii) The acceleration due to gravity, g, on the surface of Saturn is 10.4 m s^{-2}. An experiment to measure g at the equator would find a smaller value due to the planet's spin. Calculate the percentage drop in the measured value of g at Saturn's equator. [4]

Q7 The bob of a simple pendulum is made to travel in a horizontal circle, at a constant speed, as shown:

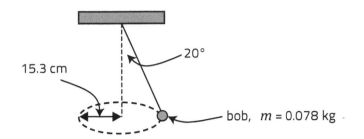

20°

15.3 cm

bob, $m = 0.078$ kg

(a) State what provides the centripetal force. [1]

(b) By considering the tension in the string, determine the speed of the bob. [5]

Q8 (a) Show that the orbital speed, v, of an object in circular orbit of radius, r, about an object of much larger mass, M, is given by $v^2 = \dfrac{GM}{r}$. [2]

(b) Dust at the visible edge of a galaxy has an orbital speed of 4700 km s^{-1}. When the distance from the centre of the galaxy is doubled, dust has a measured speed of 3400 km s^{-1}. Evaluate whether, or not, these data disprove the existence of dark matter. [3]

Question and mock answer analysis

Q&A 1

(a) Explain the difference between velocity and angular velocity for an object in circular motion at constant speed and state the relationship between the magnitude of the velocity and angular velocity. [3]

(b) The planet Wolf 1061c is in circular motion of radius 12.6×10^6 km about the star Wolf 1061 and its orbital period is 17.9 days. Calculate:

    (i) the angular velocity of the planet about the star [2]

    (ii) the orbital speed of the planet [2]

    (iii) the centripetal acceleration of the planet [2]

    (iv) the gravitational force exerted by the star Wolf 1061 on the planet Wolf 1061c given that the mass of the planet Wolf 1061c is 2.6×10^{25} kg. [2]

(c) Calculate the mass of the star Wolf 1061. [2]

(d) The star Wolf 1061 is a red dwarf with a low surface temperature and the planet Wolf 1061c has a surface value of g 60% greater than that of Earth. Jemima, a NASA spokesperson, claims that any animals living on Wolf 1061c would be physically strong but that they would not be able to tolerate UV light. Discuss to what extent Jemima might be correct. [3]

What is being asked?

Part (a) is simply a definition question in disguise. If the difference between two terms is required, you'll find that a definition of each term will suffice. Although part (b) looks novel and strange at first sight, it is a very straightforward application of circular motion equations. Part (c) is synoptic because it requires gravitational fields, which is on Component 2, but you must remember that all papers have a minimum synoptic content. The final part does not seem to be on Component 1 either and is probably this examiner's attempt at an issue question (probably trying to hit 'Consider applications and implications of science and evaluate their associated benefits and risks.').

Mark scheme

Question part		Description	AOs			Total	Skills	
			1	2	3		M	P
(a)		Velocity, v, is the rate of change of displacement [1]	3			3		
		Angular velocity, ω, is the angle, $\Delta\theta$ (in radian), swept out divided by the time, Δt [1]						
		Relationship $v = r\omega$, where r is radius (of circular motion) [1]						
(b)	(i)	Use of angle in radian divided by time [1]	1	1		2	1	
		$\omega = 4.06 \times 10^{-6}$ rad s^{-1} [1]						
	(ii)	Use of $v = r\omega$ **or** $v = \dfrac{2\pi r}{t}$ [1]	1	1		2	1	
		Correct answer (ecf on ω but needs comment if speed is greater than 3×10^8 m s^{-1}) $= 51\ 200$ m s^{-1} [1]						
	(iii)	Use of $a = \omega^2 r$ **or** $a = \dfrac{v^2}{r}$ [1]	1	1		2	1	
		Correct answer (ecf on ω, v and km) $= 0.208$ m s^{-2} [1]						
	(iv)	Realising gravitational force is the centripetal force (can be implied by answer) [1]	1	1		2	2	
		Correct answer $= ma = 5.41 \times 10^{24}$ N (ecf on a) [1]						
(c)		Rearrangement of $F = \dfrac{GMm}{r^2}$ i.e., $M = \dfrac{Fr^2}{Gm}$ [1]		2		2	2	
		Correct answer $= 4.95 \times 10^{29}$ kg [1]						

(d)	Physical strength linked to strong gravity [1]							
	Lower temperature linked to less UV [1]							
	Acceptable conclusion linked to both of the above points or to a new point, e.g. Jemima is wrong because we cannot assume life exists on this planet **or** Jemima's conclusion is correct linked to sensible strength and UV comments [1]				3	3		
Total				6	7	3	16	7

Rhodri's answers

(a) Velocity is distance over time and angular velocity is angle over time ✓ bod

The equation relating them is $v = wr$

MARKER NOTE
Rhodri's definition of velocity is poor – he really should use the word displacement and not distance. His definition of angular velocity is also poor but is far closer to that required in the MS and has been awarded the mark with bod. He has not explained the meaning of r in his equation and cannot receive the last mark.

1 mark

(b) (i) $\dfrac{360}{(17.9 \times 24 \times 3600)}$ ✗

$= 2.33 \times 10^{-4} \,°\text{s}^{-1}$ ✗

MARKER NOTE
Rhodri has forgotten to use angles in the unit of radian and cannot obtain the 1st mark. There is no ecf within a question part and so he loses the 2nd mark also.

0 marks

(ii) $v = \omega r = 2.33 \times 10^{-4} \times 12.6 \times 10^{6}$ ✓

$= 2935.8 \text{ m/s}$ ✗ no ecf

MARKER NOTE
Rhodri has used the correct equation and gains the 1st mark. He cannot gain the 2nd mark because he has forgotten to convert km to m.

1 mark

(iii) $a = \omega^2 r = (2.33 \times 10^{-4})^2 \times 12.6 \times 10^{6}$ ✓

$= 0.67 \text{ m s}^{-2}$ ✓ ecf

MARKER NOTE
Rhodri has already been penalised for not converting km to m and should not be penalised again. He has also been penalised for his incorrect angular velocity. This means that this answer, with ecf, gains full marks.

2 marks

(iv) $F = ma = 2.6 \times 10^{25} \times 0.67$ ✓

$= 1.7 \times 10^{25} \text{ N}$ ✓ ecf

MARKER NOTE
Rhodri has used the correct equation and his answer is correct with ecf and so he obtains full marks. Notice that he didn't have to state that the gravitational force was the centripetal force because this is implied in his final answer.

2 marks

(c) I don't understand why this question is on this paper. I wonder what OFQUAL will say about this Mr Examiner?

$M = \dfrac{Gm}{Fr^2}$ ✗ $= 6.4 \times 10^{-25} \text{ kg}$ ✗

MARKER NOTE
Rhodri doesn't seem to know that all papers have a synoptic content and so an examiner can ask about Component 2. Hence this question is fully in line with Ofqual requirements. Rhodri cannot gain any marks because he has rearranged the equation incorrectly and the 1st mark is for correct rearrangement.

0 marks

(d) This is definitely nothing to do with Component 1. Jemima is probably right because you'd have to be extra strong to cope with the bigger gravity ✓ and they don't sell suntan lotion on Wolf planets.

MARKER NOTE
Once again, Rhodri provides a little light relief for the examiner. He gains the 1st mark for linking extra strength to the stronger gravity. He does not gain the last mark because his UV comment, although funny, is not sensible.

1 mark

Total **7 marks /16**

Ffion's answers

(a) Velocity is the rate of change of displacement ✓
whereas angular velocity is the rate of change of angle (in rad) ✓
Also, velocity is a vector whereas angular velocity isn't.
The angular velocity is the velocity divided by the radius. ✓

> **MARKER NOTE**
> Ffion's definition of velocity is excellent and her definition of angular velocity is, arguably, better than that of the MS. Her fact about angular velocity not being a vector is actually wrong in 3D but this is well beyond the scope of A level and would not be penalised here. Her final relationship is clear with all terms named.
> **3 marks**

(b) (i) $\omega = \dfrac{\theta}{t} = \dfrac{2\pi}{17.9}$ ✓ $= 0.351$ ✗

> **MARKER NOTE**
> Although Ffion's answer is a long way from being correct, she is fortunate to obtain the 1st mark because she has used the correct unit for the angle. Her answer is incorrect because she has not converted day to s and does not merit the 2nd mark.
> **1 mark**

(ii) $v = \omega r = 0.351 \times 12.6 \times 10^6$ ✓
$= 4.4 \times 10^6$ km/s
(oops seems too large) ✗ [not enough for ecf]

> **MARKER NOTE**
> Ffion's answer is faster than the speed of light but is correct with ecf. To gain the final mark by ecf she needed to make it clear that it was greater than c.
> **1 mark**

(iii) $a = \omega^2 r = 0.351 \times 12.6 \times 10^9$ ✓ bod
$= 4.4 \times 10^9$ m s^{-2} ✗ no ecf

> **MARKER NOTE**
> Ffion is fortunate to obtain a mark here. Although she has written the correct equation, she has forgotten to square the angular velocity. The examiner has awarded the 1st mark with bod because it seems like she is attempting to use the correct equation.
> **1 mark**

(iv) The gravitational force provides the centripetal force. Hence, ✓
$F = ma = 2.6 \times 10^{25} \times 4.4 \times 10^9 = 1.1 \times 10^{35}$ N ✓ ecf

> **MARKER NOTE**
> Ffion's answer is a long way from the correct answer but each incorrect value that she has used has been punished before. She gains full marks with ecf. Notice how hard the examiners have to work to ensure that ecf is applied correctly!
> **2 marks**

(c) $M = \dfrac{Fr^2}{Gm}$ ✓
$= 1 \times 10^{34}$ kg ✗ no ecf

> **MARKER NOTE**
> Ffion obtains the 1st mark for a correct rearrangement but she cannot gain the last mark by ecf because she has forgotten to convert km into m. No ecf is available to Ffion here because this is the first time that she has made this mistake. Again, notice how difficult it is to be an examiner!
> **1 mark**

(d) Extra strength would be useful if g was 16 m s^{-2} because you would have to do more work against gravity when lifting stuff ✓. This star's blackbody spectrum definitely would give less UV because of the lower temperature ✓. However, Jemima's comments might be nonsense because the atmosphere could be carbon dioxide and sulfur dioxide like Venus's.

> **MARKER NOTE**
> Ffion's remarks about gravity and UV are to the point and well expressed. Her final comment seems to be about the impossibility of life itself and does not address the question.
> **2 marks**

Total	11 marks /16

Section 6: Vibrations

Topic summary

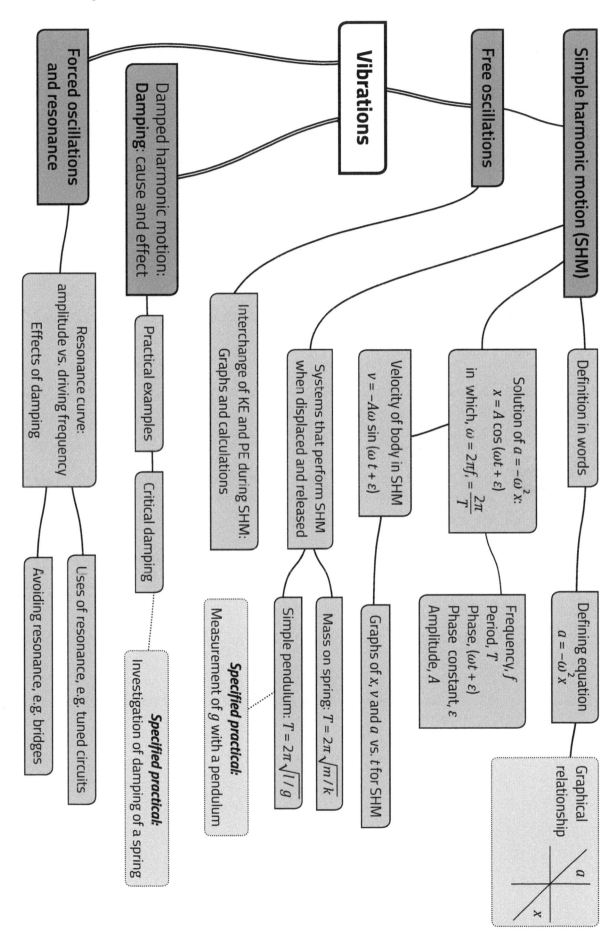

Vibrations

Simple harmonic motion (SHM)

Definition in words

Solution of $a = -\omega^2 x$:
$x = A \cos(\omega t + \varepsilon)$
in which, $\omega = 2\pi f, = \dfrac{2\pi}{T}$

Velocity of body in SHM
$v = -A\omega \sin(\omega t + \varepsilon)$

Defining equation
$a = -\omega^2 x$

Graphical relationship

Frequency, f
Period, T
Phase, $(\omega t + \varepsilon)$
Phase constant, ε
Amplitude, A

Graphs of x, v and a vs. t for SHM

Free oscillations

Systems that perform SHM when displaced and released

Simple pendulum: $T = 2\pi \sqrt{l/g}$

Mass on spring: $T = 2\pi \sqrt{m/k}$

Specified practical:
Measurement of g with a pendulum

Interchange of KE and PE during SHM:
Graphs and calculations

**Damped harmonic motion:
Damping: cause and effect**

Practical examples

Critical damping

Specified practical:
Investigation of damping of a spring

Forced oscillations and resonance

Resonance curve:
amplitude vs. driving frequency

Effects of damping

Avoiding resonance, e.g. bridges

Uses of resonance, e.g. tuned circuits

Q1   A displacement–time graph is given on the left for a body performing SHM.

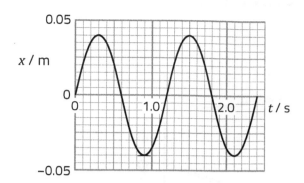 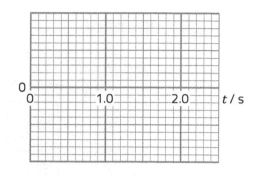

(a)   For this oscillation, determine the values of A, ω and ε in the equation

$$x = A \cos (\omega t + \varepsilon).$$

(i) A .... [1]

(ii) ω .... [2]

(iii) ε .... [1]

(b)   Using the grid on the right above, sketch a velocity–time graph for the body, providing a vertical scale. Use the space below for working.   [4]

Q2   A metal ball attached to a spring whose other end is fixed is given a displacement $x = +0.140$ m from its equilibrium position and released at time $t = 0$. It performs SHM of period 0.800 s. Find:

(a)   (i)   The ball's displacement at $t = 0.50$ s.   [3]

....

....

....

....

....

(ii)   The first and second times at which the ball's displacement is +0.070 m. [A rough sketch graph may help.]   [3]

....

....

....

....

(b) (i) · The ball's velocity at $t = 0.50$ s. [3]

....

....

....

....

(ii) The first and second times at which the ball's velocity is +0.55 m s^{-1}. [A rough sketch graph may help.] [3]

....

....

....

....

....

Q3 The diagram shows a system that performs SHM of frequency 0.40 Hz, in a horizontal plane, when displaced from its equilibrium position and released.

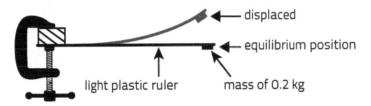

displaced

equilibrium position

light plastic ruler   mass of 0.2 kg

The acceleration, a, of the mass is related to its displacement, x, by the equation:

$$a = \frac{k}{m}x$$

(a) On what part of the system does the value of k depend? [1]

....

(b) Calculate the value of k. [3]

....

....

....

....

(c) Sketch a velocity–time (v–t) graph for the first two cycles of the mass's motion, if it is released from rest with displacement $x = +0.050$ m at time $t = 0$. Mark significant values of t and the value of maximum v on the axes. [3]

Space for calculation:

v / m s^{-1}

0

t / s

Q4 On the Moon, 100 small oscillations of a simple pendulum take 240 s. On the Earth, 100 small oscillations of the same pendulum take 100 s. Calculate a value for g on the Moon. [3]

Q5 A load of mass 200 g hangs from a spring of stiffness 40 N m^{-1}. The top end of the spring is firmly clamped. The load is pulled down 20 mm below its equilibrium position and then released.

(a) Calculate the time the load takes to reach its highest point. [2]

(b) Fergus says that if the load had been pulled down 30 mm it would have taken a shorter time to reach its highest point, because it would experience a larger resultant upward force. Evaluate Fergus's claim. [3]

Q6 A small object is placed on a spring and the spring extends by a length l. A pendulum is then made of length l (exactly the same as the extension of the spring) and placed to oscillate next to the object on the spring. Davinder notices that both the pendulum and the object on the spring oscillate with exactly the same frequency. Davinder states that this is a complete coincidence. Discuss to what extent Davinder is correct. [5]

Q7 A simple pendulum has a length of 1.00 m. The mass of its bob is 100 g. The pendulum is displaced from the vertical by an angle of 11.5°. The horizontal displacement of the pendulum bob is 0.20 m.

(a) Show that the potential energy of the bob (relative to its lowest point) is approximately 0.02 J. A sketch diagram might help. [3]

....

....

....

....

....

....

(b) The pendulum is released. A graph of (horizontal) displacement, x, against time is given.

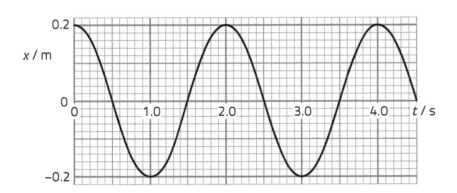

On the grids below sketch graphs of:

(i) The pendulum's potential energy, E_p, against time. [3]

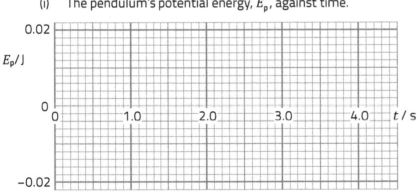

(ii) The pendulum's kinetic energy, E_k, against time. [2]

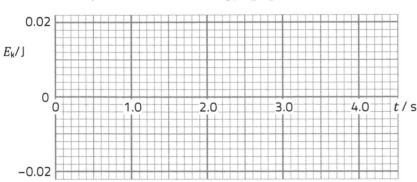

Q8 A velocity–time graph is given for a disc of mass 0.24 kg attached to a spring and oscillating in air.

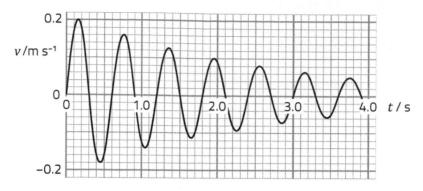

(a) Explain, in terms of forces, why the oscillations are damped. [2]

....

....

....

(b) Sophie believes that the peak values of velocity decrease *exponentially* with time. Evaluate her claim. [3]

....

....

....

....

....

(c) Determine the percentage of the disc's kinetic energy that is dissipated over the three cycles between the first positive peak and the fourth positive peak. [2]

....

....

....

Q9 (a) State what is meant by *critical damping*. [2]

....

....

....

(b) State one use for critical damping, explaining why lighter damping would not be as suitable. [3]

....

....

....

....

Q10 (a) Define *forced oscillations*. [2]

....

....

....

(b) The diagram shows apparatus that can be used for investigating forced oscillations of a simple pendulum. You are not expected to have seen this apparatus before. [2]

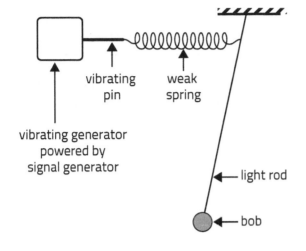 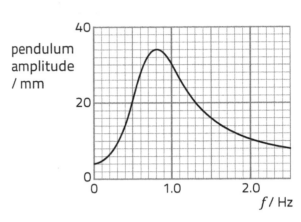

(i) Identify the *driving force* on the pendulum. [1]

....

(ii) Calculate a value for the length of the pendulum, explaining your reasoning. [4]

....

....

....

....

....

....

Question and mock answer analysis

Q&A 1

(a) Define *simple harmonic motion (SHM)*. [2]

(b) The diagram shows a system that can perform SHM.

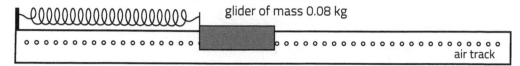

spring (of equal stiffness in compression and extension)

glider of mass 0.08 kg

air track

The glider is displaced along the air track from its equilibrium position and released at time $t = 0$. A displacement–time graph is given on the left tor the glider's motion after release.

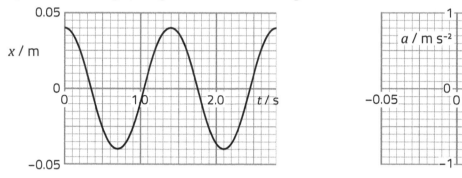

(i) Give the values of the amplitude **and** the period. [2]

(ii) On the grid provided to the right above, draw a graph of acceleration, a, against displacement, x, for the glider. [Space for working is given.] [4]

(iii) Calculate the maximum value of the glider's kinetic energy. [2]

(iv) Ahmed claims that kinetic and potential energy vary at a frequency of 1.43 Hz. Evaluate this claim. [3]

What is being asked

You will have seen the oscillations of a mass hanging from a spring. The apparatus in the question is clearly harder to set up (and a lot more expensive!) but the physics is actually rather easier!

(a) A definition that sets the scene.

(b) (i) No catches, but read the scales carefully!

(ii) Even if you haven't previously met the graph that's asked for, a little thought should tell you the shape. Note that the word 'draw' is used, rather than 'sketch', so your line has to go through the right points.

(iii) This is intended to be a straightforward calculation that shifts the thrust of the question towards the energy aspect of SHM, in preparation for...

(iv) 'Evaluate' is a broad hint that AO3 is being tested. You have to figure out how to approach the evaluation. You could start by mustering what you know about energy in SHM, or you could first try to make sense of the 1.43 Hz.

Mark scheme

Question part		Description	AOs			Total	Skills	
			1	2	3		M	P
(a)		Acceleration proportional to displacement from equilibrium [1] and in opposite direction **or** directed towards equilibrium [1]	2			2		
(b)	(i)	[Amplitude =] 0.040 m [1] [Period =] 1.40 s [1]		2		2		
	(ii)	$\omega = 4.49$ s^{-1} or by implication [1] $\omega^2 = 20.1$ s^{-2} or by implication [1] Line runs between $x = -0.04$ m and $+0.04$ m [1] and between $a = +0.8$ m s^{-2} and -0.8 m s^{-2} with negative gradient [1]		4		4	4	
	(iii)	$v_{max} = 0.040 \times 4.49$ [1] [$= 0.180$ m s^{-2}] or by implication $E_{k\,max} = 1.29$ mJ **unit** [1]		2		2	2	
	(iv)	1.43 Hz shown to be twice the frequency of the displacement variation [1] Any argument that KE varies at this 'double' frequency, e.g. maxima at each pass through equilibrium [1] Any argument that PE varies at this 'double' freq., e.g. maxima at max spring extension and max spring compression [1]			3	3	1	
Total			2	8	3	13	7	0

Rhodri's answers

(a) Acceleration is proportional to displacement. ✓ X

MARKER NOTE
Rhodri has left out the statement about the direction. It matters!

1 mark

(b) (i) 0.035 m X   1.40 s ✓

MARKER NOTE
A mistake in reading the amplitude but the period is correct.

1 mark

(ii) $\omega = \dfrac{2\pi}{T} = \dfrac{2\pi}{1.4} = 4.49$ ✓

max acc $= 4.49 \times 0.035 = 0.16$ X

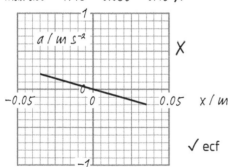

✓ ecf

MARKER NOTE
Rhodri gains the 1st mark, for calculating ω. But he hasn't squared it and loses both the 2nd and 4th marks. With ecf on A from (b)(i) he gains the 3rd mark.

2 marks

(iii) $v = -A\omega\sin\omega t$

Maximum is when $t = 1.05$ s

$v = -0.35 \times 4.49 \sin(4.49 \times 1.05)$ X

$= -0.129$

Max KE $= \dfrac{1}{2} \times 0.08 \times (-0.129)^2$

$= 6.7 \times 10^{-6}$ J X [no ecf]

MARKER NOTE
Rhodri hasn't seen that the extreme values of the sine function are ±1, so $v_{max} = A\omega$. Instead he's correctly identified the time for maximum v, but has put ωt in radians into a calculator set to degrees. A common mistake, but a costly one.

0 marks

(iv) The frequency is 1/T = 0.714 Hz ✓

Ahmed's figure of 1.43 Hz is double this, so Ahmed is wrong.

MARKER NOTE
Rhodri has clearly gained the first mark, but hasn't realised that the variation in energy *should be* at this double frequency.

1 mark

Total   **5 marks /13**

Ffion's answers

(a) The acceleration is proportional to the distance from a fixed point ✓ and is directed towards that point. ✓

MARKER NOTE
Ffion's statement is correct, though it is safer to use 'displacement' rather than 'distance' (as x in the equation is displacement).

2 marks

(b) (i) 0.04 m ✓ 1.4 s ✓

MARKER NOTE
Both readings correct. It would have been nice to see an extra significant figure given for both values but this was not penalised.

2 marks

(ii) $a_{max} = \omega^2 A = \left(\dfrac{2\pi}{1.4}\right)^2 0.04 = 0.81$ m s^{-2} ✓✓

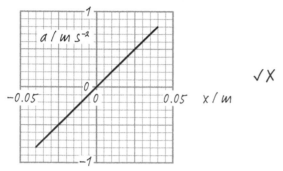

✓ X

MARKER NOTE
Ffion has done everything correctly except for forgetting the minus sign, so her graph gradient is positive instead of negative. She loses the 4th mark.

3 marks

(iii) $v_{max} = A\omega = 0.040 \times \dfrac{2\pi}{1.4} = 0.1795$ m s^{-1} ✓

$(KE)_{max} = \dfrac{1}{2}mv_{max}^2 = \dfrac{1}{2}0.08 \times 0.1795^2$
$= 1.29 \times 10^{-3}$ J ✓

MARKER NOTE
Ffion has used $v_{max} = A\omega$, and has done the calculation correctly.

2 marks

(iv) PE is maximum at each extreme of the displacement and KE is a maximum in the middle, which ever way the glider is going. ✓ So the energy is transferred twice as often as the oscillation frequency, and Ahmed is right. ✓

MARKER NOTE
Ffion understands what is going on, but hasn't actually *shown* that 1.43 Hz is twice the ordinary oscillation frequency, losing the 1st mark.

2 marks

| Total | 11 marks /13 |

Section 7: Kinetic theory

Topic summary

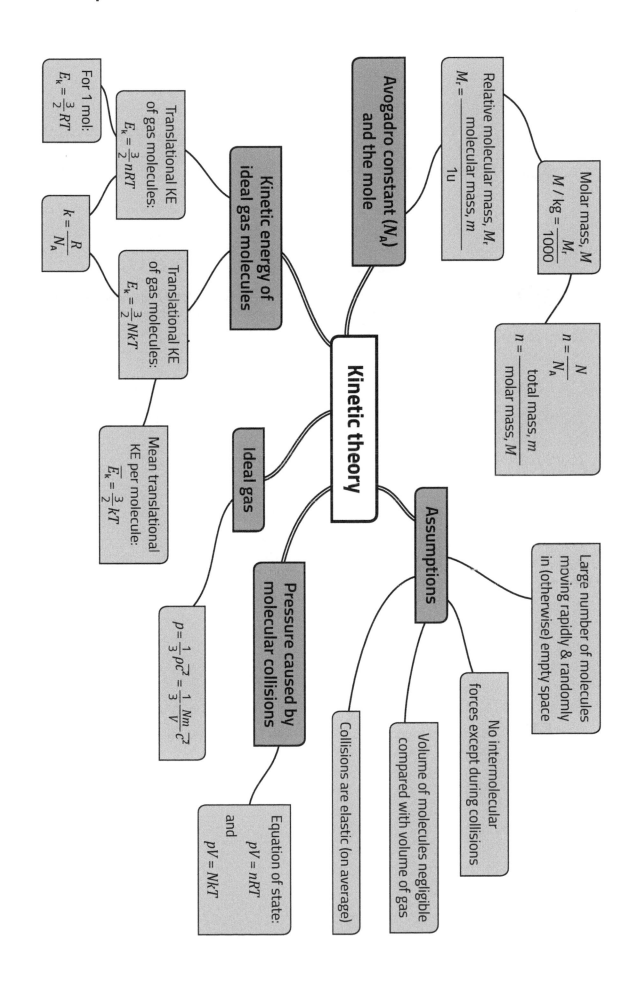

Kinetic theory

Avogadro constant (N_A) and the mole

Relative molecular mass, M_r
$$M_r = \frac{\text{molecular mass, } m}{1u}$$

Molar mass, M
$$M / \text{kg} = \frac{M_r}{1000}$$

$$n = \frac{N}{N_A}$$
$$n = \frac{\text{total mass, } m}{\text{molar mass, } M}$$

Kinetic energy of ideal gas molecules

Translational KE of gas molecules:
$$E_k = \frac{3}{2} nRT$$

For 1 mol:
$$E_k = \frac{3}{2} RT$$

$$k = \frac{R}{N_A}$$

Translational KE of gas molecules:
$$E_k = \frac{3}{2} NkT$$

Mean translational KE per molecule:
$$\overline{E_k} = \frac{3}{2} kT$$

Ideal gas

$$p = \frac{1}{3}\rho \overline{c^2} = \frac{1}{3}\frac{Nm}{V}\overline{c^2}$$

Pressure caused by molecular collisions

Equation of state:
$$pV = nRT$$
and
$$pV = NkT$$

Assumptions

Large number of molecules moving rapidly & randomly in (otherwise) empty space

No intermolecular forces except during collisions

Volume of molecules negligible compared with volume of gas

Collisions are elastic (on average)

Q1 State the assumptions of the kinetic theory of gases. [4]

Q2 Define the Avogadro constant and the mole. [2]

Q3 Explain, in terms of movement of molecules and Newton's laws, how a gas exerts pressure on its container walls and how this pressure varies with temperature. [6 QER]

Q4 Use the equations $pV = nRT$ and $pV = NkT$ to derive the relationship $k = \dfrac{R}{N_A}$. Make clear what the symbols n, N and N_A represent. [3]

Q5 Use the equations $pV = \frac{1}{3}Nm\overline{c^2}$ and $pV = nRT$ to show that the translational kinetic energy, U, of n mol of monatomic gas is given by $U = \frac{3}{2}nRT$. [3]

....

....

....

....

....

Q6 The explosion inside a car-safety airbag produces 3.0 mol of nitrogen gas (relative molecular mass = 28). The pressure inside the airbag is 140 kPa and the rms speed of the nitrogen molecules is 550 m s^{-1}. Calculate the volume of the airbag. [4]

....

....

....

....

....

....

....

Q7 A meteorological balloon is released from ground level. The helium in the balloon has an initial volume of 0.89 m^3 and a temperature of 298 K. The pressure at ground level is 102 kPa.

loose envelope of balloon

(a) Calculate the number of molecules of helium in the balloon. [3]

....

....

....

....

....

(b) Calculate the rms speed of the helium molecules (the mass of a helium molecule is 6.64×10^{-27} kg). [2]

....

....

(c) The balloon rises to a height where the pressure is 23 kPa and the temperature is 232 K. Calculate the new volume of the balloon stating any assumption that you make. [3]

....

....

....

....

Q8 Air is contained in two separate containers connected by a narrow tube fitted with a tap. The air in both containers is in thermal equilibrium with the surroundings, the temperature of which is 293 K.

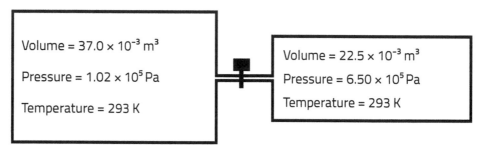

Volume = 37.0×10^{-3} m³

Pressure = 1.02×10^5 Pa

Temperature = 293 K

Volume = 22.5×10^{-3} m³

Pressure = 6.50×10^5 Pa

Temperature = 293 K

The tap is opened and air flows from the right container to the left until the pressures in the two containers are equal and the containers are in thermal equilibrium with the surroundings.

(a) Calculate the final pressure in the containers. [5]

....

....

....

....

....

....

....

(b) Before thermal equilibrium is reached, Tudor claims that the right container will cool and the left container will become warmer. Discuss whether Tudor is correct. [3]

....

....

....

....

....

Q9   Air of density 1.35 kg m^{-3} and temperature 293 K, with a pressure of 112 kPa, is trapped in a Cola bottle of volume 1.5×10^{-3} m^3.

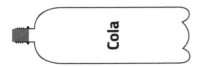

(a)   Calculate the rms speed of the air molecules.   [3]

....

....

....

....

(b)   Calculate the mass of 1 mol of air.   [3]

....

....

....

....

(c)   (i)   Air is gradually pumped into the bottle until the bottle explodes when the pressure is 935 kPa and the temperature of the air is 320 K. Calculate the mass of air inside the bottle when this occurs, stating any assumption you make.   [3]

....

....

....

....

(ii)   Explain why the temperature inside the bottle increases as air is pumped in.   [2]

....

....

Component 1 Practice questions

Question and mock answer analysis

Q&A 1

(a) A sample of an ideal gas, of amount 0.078 mol, at a temperature of 157 K, has a volume of $1.45 \times 10^{-3}\,\text{m}^3$.

(i) Calculate the pressure of the gas. [2]

(ii) Explain what is meant by 0.078 mol. [1]

(iii) The density of the gas is 4.52 kg m^{-3}. Calculate the rms speed of the molecules and the molar mass of the gas. [5]

(iv) Gwesyn makes two statements about the gas molecules. In each case determine whether the statement is correct:

Statement 1: 'Halving the mass of each molecule would halve the mean kinetic energy of the molecules (for a given temperature).'

Statement 2: 'Doubling the temperature would double the rms speed of the molecules.' [5]

(b)

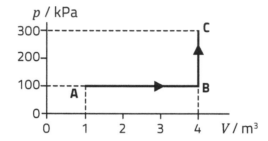

A different sample of a monatomic gas is taken from point A to point C on the p–V diagram. Use calculations to compare:

(i) the temperatures at A, B and C, and [2]

(ii) the ways in which energy is transferred between the gas and its surroundings during the two stages, AB and BC. [3]

What is being asked?

Part (a)(i) is a simple, one-step calculation; the 1st mark is AO1 but the 2nd mark requiring a calculation is AO2. Part (ii) is related to the definition of the mole (and hence is an AO1 mark). The next two calculations in part (iii) are a little trickier, with the second requiring a two-step calculation. There are a couple of different ways of obtaining the correct answers and these are AO2 marks. Part (iv) is a standard AO3 question in which you have to evaluate statements using physics – with several ways of obtaining the answer. Part (b) overlaps with the Thermal physics area of study, which is quite common as the two have many ideas in common.

Mark scheme

Question part		Description	AOs			Total	Skills	
			1	2	3		M	P
(a)	(i)	Substitution into $pV = nRT$ [1] Answer = 70 200 Pa [1]	1	1		2	1	
	(ii)	$N / N_A = 0.078$ or equivalent, e.g. 4.7×10^{22} particles [1]	1			1		
	(iii)	Rearrangement of $p = \frac{1}{3}\rho \overline{c^2}$, i.e. $\overline{c^2} = \dfrac{3p}{\rho}$ (minimum) **or** alternative, e.g. $\frac{3}{2}kT$ and obtaining mass m [1] rms speed = 216 m s^{-1} (correct answer only) [1] Total mass = ρV = 6.55 g **or** $m = 1.39 \times 10^{-25}$ kg [1] Dividing by the number of moles OR 1u (not both) [1] Answer = 84 g mol^{-1} (or 0.084 kg mol^{-1}) ***unit*** [1]		5		5	5	

	(iv)	**Statement 1**: Kinetic energy only depends on temperature or equation, e.g. $\frac{1}{2}m\overline{c^2} = \frac{3}{2}kT$ [1] Hence Gwesyn is wrong, i.e. correct conclusion linked to correct physics [1] **Statement 2**: KE $\propto T$ or equivalent, e.g. $\frac{1}{2}m\overline{c^2} = \frac{3}{2}kT$ [1] $c_{rms} \propto \sqrt{T}$ or equiv **or** calculation $\longrightarrow c_{rms} = 305$ m s^{-1} [1] c_{rms} increases by factor $\sqrt{2}$ **or** T must increase by $\times 4$ for it to be true **or** equivalent [1]			5	5	2	
(b)	(i)	Using $pV = nRT$ or $pV \propto T$ [1] $T_C:T_B:T_A = 12:4:1$ [1]			2	2		
	(ii)	At least one of: U(A) = 150 kJ, U(B) = 600 kJ and U(C) = 1800 kJ **or** ΔU(AB) = 450 kJ, ΔU(BC) = 1200 kJ [1] W(BC) = 0 **and** W(AB) = 300 kJ [1] Q(AB) = 750 kJ; Q(BC) = 1200 kJ [1]			3	3		
Total			2	6	10	18	8	

Rhodri's answers

(a) (i) $p = \dfrac{nRT}{V} = \dfrac{0.078 \times 8.31 \times 157}{1.45 \times 10^{-3}}$ ✓

$= 70\,100$ Pa ✓ [bod]

> **MARKER NOTE**
> Rhodri's working is perfectly correct. He has rounded the pressure (70.18 kPa) incorrectly but this is not always penalised (except in practical questions) – it is correct to 2 sf anyway.
> **2 marks**

(ii) Hah, a doddle, you'll have $0.078 \times N_A$ molecules ✓

> **MARKER NOTE**
> Rhodri's answer is equivalent to the mark scheme. Irrelevant comments are ignored unless they are rude or imply a safeguarding issue. **1 mark**

(iii) $\overline{c^2} = \dfrac{3p}{\rho}$ ✓ $= 46\,500$ m s^{-1}

$M = \rho V = 0.006554$ ✓

$Mm = 0.006554 / (0.078 \times 6.02 \times 10^{23})$

$Mm = 1.396 \times 10^{-25}$

> **MARKER NOTE**
> Rhodri has made a common mistake. He has calculated the mean square speed and forgotten to take the square root.
> Rhodri's 2nd answer is also incorrect because he has calculated the mass of a molecule rather than a mole. Mixing up the various masses is a common mistake. Note that Rhodri cannot obtain the penultimate mark because he has divided by both 0.078 and N_A. **2 marks**

(iv) Since KE $= m\overline{c^2}$

It's obvious that KE is proportional to mass and Gwesyn is right ✗

> **MARKER NOTE**
> Rhodri has fallen into the trap here and has forgotten that lighter particles will have greater rms speeds at the same temperature. He has simply looked at the kinetic energy formula and come to the wrong conclusion. **0 marks**

Using $1/2 m\overline{c^2} = 3/2 kT$ ✓

So $\overline{c^2}$ is proportional to T

And Gwesyn is correct (any chance of more difficult questions next time please 😊)

> **MARKER NOTE**
> Rhodri gains the 1st mark because he realises that $\frac{1}{2}m\overline{c^2} = \frac{3}{2}kT$ is a suitable starting point. In spite of his smugness, he then makes the same mistake as in part (c) – considering the mean square speed and not the rms – an error of <u>physics</u>, so no ecf will be awarded here. **1 mark**

(b) (i) $pV = nRT$ so $T = \dfrac{pV}{nR}$ ✓

But we don't know n so we can't work out the temperature – not enough information!

> **MARKER NOTE**
> Rhodri gains a mark because $pV = nRT$ is a good starting point. He doesn't realise the significance of the instruction to <u>compare</u> the temperatures, e.g. finding the ratio, so he cannot access the 2nd mark. **1 mark**

(ii) A to B: W = area under graph

$= 3 \times 100 \times 10^3$

$= 300\,000$ J

B to C: W = 0 (volume constant) ✓

> **MARKER NOTE**
> The only mark that Rhodri accesses is the one for the two values of W. He appears not to consider Q at all, probably because he has no idea how to calculate ΔU – which is similar to his difficulty with part (i). **1 mark**

> **Total** **8 marks /18**

Ffion's answers

(a) (i) $p = \dfrac{nRT}{V} = 70$ kPa ✓✓

MARKER NOTE
Ffion has rearranged the equation correctly and has obtained the correct answer for full marks. Ffion has rounded correctly to 2 sf here which is ideal since the number of moles is only to 2 sf. Note that sfs and rounding are mainly penalised in the practical questions.   **2 marks**

(ii) The no. of particles is 0.078 × the no. of particles in 12 g of carbon-12 ✓

MARKER NOTE
Ffion's answer is equivalent to Rhodri's but she has also provided an (unnecessary) out-of-date definition of a mole (it is now defined as $6.022\,140\,76 \times 10^{23}$ particles but this happened after the Terms & definitions booklet was written).   **1 mark**

(iii) $\frac{1}{2}m\overline{c^2} = \frac{3}{2}kT$: First calculate m ✓

$M = \rho V = 4.52 \times 0.00145 = 0.006554$ kg

$m = \dfrac{6.554 \times 10^{-3}}{0.078 \times 6.02 \times 10^{23}} = 1.396 \times 10^{-25}$ kg

$\overline{c^2} = \dfrac{3kT}{m}$, so $c_{rms} = 216$ m s^{-1} ✓

Molar mass $= \dfrac{1.396 \times 10^{-25}}{1.66 \times 10^{-27}} = 87$ ✓✓ unit

MARKER NOTE
Ffion's answer is sound but she has made life difficult for herself by taking the alternative route to finding the rms speed as well as an alternative method of finding the molar mass. Also the two calculations are interwoven – a problem for the examiner! Her only omission is the final unit for the molar mass – and the other method, i.e. dividing the total mass by the number of moles makes this a less likely mistake.   **4 marks**

(iv) Using $\frac{1}{2}mc^2 = \frac{3}{2}kT$ again ✓

The right-hand side stays the same (same temp) so the left-hand side stays the same and so Gwesyn is wrong ✓

MARKER NOTE
Ffion's answer has made a nice (concise) answer which is fully correct – gaining full marks.   **2 marks**

Using $\frac{1}{2}m\overline{c^2} = \frac{3}{2}kT$ ✓

If $c \times 2$ then $c^2 \times 4$ ✓ So you would need to quadruple the temperature to double the speed and Gwesyn is wrong again. ✓

MARKER NOTE
Ffion has chosen to double the (rms) speed and show that the temperature would be multiplied by 4. An equally valid way would be to double T and show that the speed is multiplied by $\sqrt{2}$. Anyway – a perfect answer and full marks.   **3 marks**

(b) (i) $pV \propto T$.

Point A: $pV = 100k \times 1 = 100\,000$ ✓

Point B: $pV = 400\,000$

Point C: $pV = 1\,200\,000$

∴ $T_C = 12 \times T_A$ and $T_B = 4 \times T_A$ ✓

MARKER NOTE
Ffion uses the relationship $pV \propto T$ and makes no attempt to calculate the actual temperatures but is able to find the factors by which B and C are above the temperature of A – picking up both marks.   **2 marks**

(ii) A ⟶ B: $W = 100 \times (4 - 1) = 300$ kJ

$U = \frac{3}{2}nRT = \frac{3}{2}pV$

∴ $\Delta U = \frac{3}{2} \times 100 \times (4 - 1) = 450$ kJ ✓

∴ $Q = \Delta U + W = 750$ kJ

B ⟶ C: $W = 0$ ✓

$\Delta U = \frac{3}{2} \times (300 - 100) \times 4 = 800$ kJ ✗

∴ $Q = 1200$ kJ

MARKER NOTE
Ffion's answer is almost perfect. She doesn't calculate the internal energy at any of A, B or C but calculates the ΔU(AB) correctly (1st mark). W for the two stages gains her the 2nd mark. She cannot obtain the third mark because she makes a mistake in calculating ΔU(BC) – forgetting to use the factor of $\frac{3}{2}$ **2 marks**

Total   **16 marks /18**

Section 8: Thermal physics

Topic summary

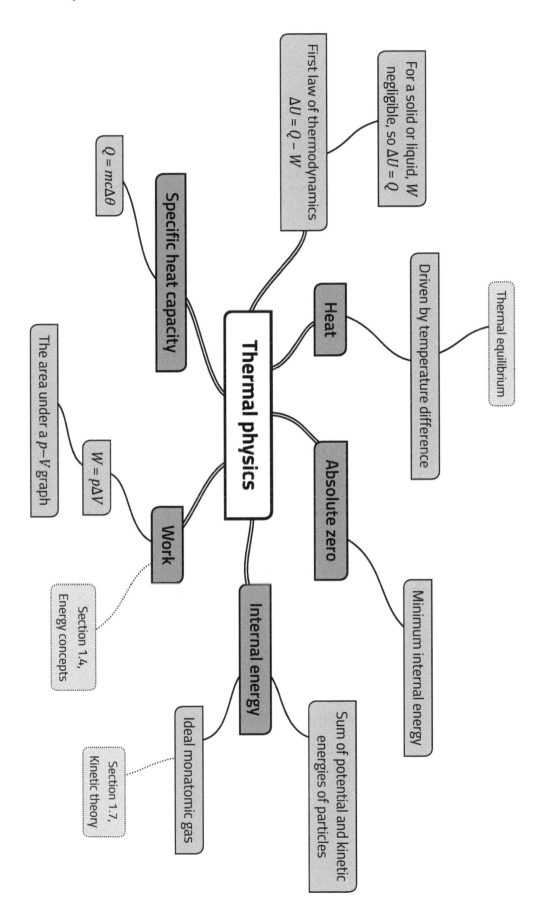

Thermal physics

Specific heat capacity
- $Q = mc\Delta\theta$

First law of thermodynamics $\Delta U = Q - W$
- For a solid or liquid, W negligible, so $\Delta U = Q$

Heat
- Driven by temperature difference
  - Thermal equilibrium

Absolute zero
- Minimum internal energy

Work
- $W = p\Delta V$
  - The area under a p–V graph
- Section 1.4, Energy concepts

Internal energy
- Ideal monatomic gas
  - Section 1.7, Kinetic theory
- Sum of potential and kinetic energies of particles

Q1 State what is meant by the *internal energy of a system*. [2]

Q2 Explain the significance of *absolute zero* with regards to internal energy. [2]

Q3 (a) Explain why the internal energy of an ideal gas is different from that of systems in general. [2]

(b) Calculate the internal energy of 30 g of neon gas at a temperature of 26.85 °C (the molar mass of neon is 20 g). [2]

Q4 Explain what is meant by the term *heat*. [2]

Q5 Two systems in thermal contact are in *thermal equilibrium*. State what *thermal equilibrium* means in terms of *heat* and *temperature*. [2]

Q6 (a) The first law of thermodynamics can be written in the form:

$$\Delta U = Q - W$$

Explain the meaning of each term in the equation and how this equation represents conservation of energy. [3]

(b)   Explain why the first law of thermodynamics reduces to $\Delta U = Q$ for a solid or liquid.   [2]

....

....

....

Q7   Define the *specific heat capacity* of a substance.   [2]

....

....

....

Q8   A group of students uses the following apparatus to investigate how the pressure of a sample of air varies with temperature, at constant volume. Describe how they could use the apparatus to obtain an estimate of absolute zero.   [6 QER]

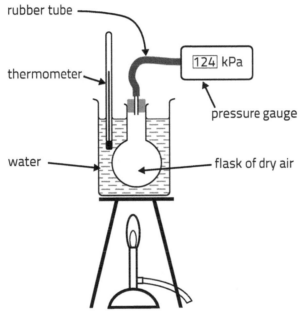

....

....

....

....

....

....

....

....

....

....

....

....

....

....

Q9 A gas is expanded quickly so that no heat is transferred to the gas. Explain why the temperature of the gas decreases. [3]

....

....

....

....

....

....

Q10 (a) The volume of a gas is increased by $2.7 \times 10^{-3}\,\text{m}^3$ at a constant pressure of $1.42 \times 10^5\,\text{Pa}$. Calculate the work done by the gas. [2]

....

....

....

(b) The same gas is then compressed from this increased volume at the constant pressure of $1.42 \times 10^5\,\text{Pa}$ to a final volume which is $1.5 \times 10^{-3}\,\text{m}^3$ less than the original volume at the start of part (a). Calculate the work done by the gas. [3]

....

....

....

....

....

Q11 Carrots of mass 0.700 kg and specific heat capacity $1880\,\text{J}\,\text{kg}^{-1}\,\text{K}^{-1}$ at a temperature of 20 °C are placed in 1.2 kg of boiling water. The specific heat capacity of water is $4210\,\text{J}\,\text{kg}^{-1}\,\text{K}^{-1}$.

Calculate the equilibrium temperature of the carrots and water, stating any assumptions that you make. [5]

....

....

....

....

....

....

....

....

Q12 A sample of gas is taken around a closed cycle ABCA as shown in the p–V diagram.

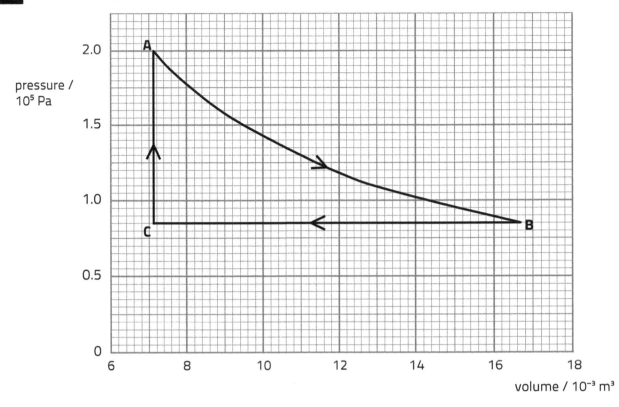

pressure /
10^5 Pa

volume / 10^{-3} m³

(a) Calculate the work done by the gas:

   (i) for process CA; [1]

       ....

   (ii) for process BC; [2]

       ....

       ....

       ....

   (iii) for process AB. [3]

       ....

       ....

       ....

       ....

       ....

(b) Charlie claims that process AB is an isotherm, i.e. it occurs at a constant temperature. Discuss why Charlie is correct. [4]

....

....

....

....

....

....

....

(c) Complete the following table with the correct data: [6]

	AB	BC	CA	ABCA
ΔU / J	0			
Q / J				
W / J	(a)(iii)	(a)(ii)	(a)(i)	

Space for calculations:

Q13 Tegfryn carries out an experiment to measure the specific heat capacity of aluminium using the standard apparatus shown below.

He uses no insulation and the temperature of the block starts from room temperature of 20 °C. He also measures the following values:

Mass of aluminium block = 1.000 kg, pd = 12.00 V, current = 4.20 A.

He checks the internet and finds that the specific heat capacity of aluminium is 900 J kg^{-1} K^{-1}.

Plot a graph of the expected results on the grid below. [6]

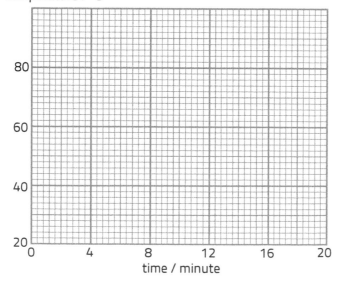

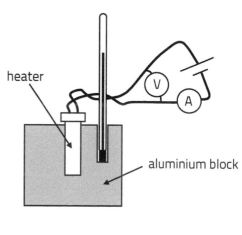

Space for calculations:

Q14 (a) Show that the equation for the work, W, done by an expanding gas

$$W = p\,\Delta V$$

is correct as far as units (or dimensions) are concerned. [2]

(b) A syringe with its outlet blocked contains 110×10^{-6} m³ of argon (a monatomic gas) at a temperature of 20 °C. The syringe is immersed in boiling water at 100 °C. The argon expands to a volume of 140×10^{-6} m³, by moving the piston of the syringe. The pressure is 100 kPa throughout.

(i) Verify that a negligible amount of gas escapes during the expansion. [4]

(ii) **Giving your reasoning clearly**, show that 7.5 J of heat enters the gas during the expansion. [5]

(iii) Lucia uses the data in this question to reach the following conclusion: 'The heat needed to raise the temperature of 1.0 mol of argon gas by 1.0 K is 21 J.' Examine to what extent her statement is justified. [4]

Question and mock answer analysis

Q&A 1

(a) The first law of thermodynamics can be written in the form:

$$\Delta U = Q - W$$

State the meaning of the three terms: ΔU, Q and W. [3]

(b) An ideal monatomic gas goes through the cycle ABCA shown in the graph:

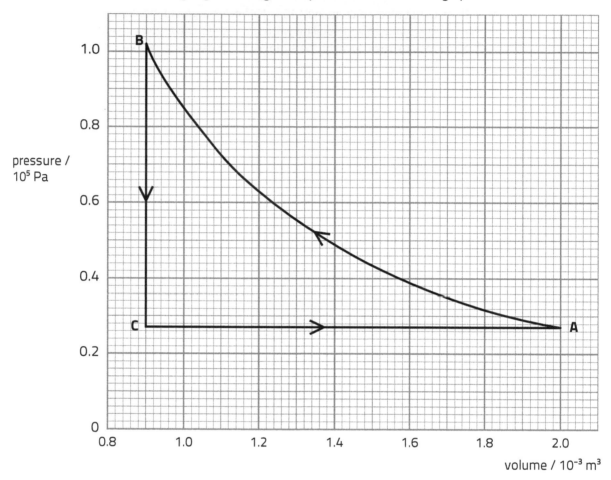

(i) The temperature at A is 293 K. Calculate the number of moles of gas. [2]

(ii) Calculate the temperatures at points B and C. [3]

(iii) Calculate the change in internal energy between A and B. [2]

(iv) Show that the work done by the gas during part AB is approximately −60 J. [3]

(v) Teilo states that, during process AB, no heat is transferred. Determine to what extent Teilo is correct. [2]

(vi) Calculate the values of ΔU, Q and W for the whole cycle ABCA. [4]

What is being asked?

Part (a) is the gentle introduction with three very easy AO1 marks. Part (b) is mainly analysis of a gas cycle. Parts (i) & (ii) are related to the previous topic (kinetic theory) but you can expect a mix 'n' match of these two topics often. Nonetheless, these are not difficult calculations – they involve one equation only ($pV = nRT$). Part (iii) is based on this topic but is also a simple one-step calculation. Part (iv) is a little more complicated and requires an area below the curve to be approximated. Whereas (i)–(iv) are mainly AO2 skills, the wording of (v) is typical of AO3 marks. There are many ways to complete part (v), as is usually the case with these questions. The final part tests AO2 skills and requires a good understanding of both a closed cycle and the first law of thermodynamics.

Mark scheme

Question part		Description	AOs 1	AOs 2	AOs 3	Total	Skills M	Skills P
(a)		ΔU, change in internal energy [1] Q, heat supplied to gas/system [1] W, work done by gas/system [1]	3			3		
(b)	(i)	Use of $pV = nRT$ [1] $n = 0.0222$ mol [1]	1	1		2	1	
	(ii)	Either use of $pV = nRT$ or pV/T = constant [1] $T_B = 498$ K [1] $T_C = 132$ K [1]	1	2		3	2	
	(iii)	Use of $U = \frac{3}{2}nRT$ or $\frac{3}{2}pV$ [1] Answer = 57 J (ecf) [1]	1	1		2	1	
	(iv)	Application of work = area under curve [1] Realisation of work done on gas or –ve sign linked to compression [1] Accurate answer shown [accept 52 J to 64 J] [1]		3		3	3	
	(v)	Application of first law, e.g. $\Delta U = Q - W$ $60 = 0 - (-60)$ OR $Q = \Delta U + W = 60 - 60 = 0$ [1] valid conclusion in light of first law (ecf) [1]			2	2	1	
	(vi)	$\Delta U = 0$ (start & end temps are equal) [1] Area for CA calculated or implied e.g. $\Delta V \times p = 1.1 \times 10^{-3} \times 0.27 \times 10^5$ (29.7 J) [1] hence $W = -30$ J [1] $Q = W$ [1]		4		4	2	
Total			6	11	2	19	10	

Rhodri's answers

(a) ΔU is internal energy **X** (not enough)

Q is heat going in or out of the gas **X**

W is work done on or by the gas **X**

MARKER NOTE
Each of Rhodri's answers falls short of the mark scheme and he gains no marks at all. He is not far from gaining 3 marks but each statement is slightly wrong or ambiguous.
0 marks

(b) (i) $n = \dfrac{pV}{RT} = \dfrac{0.27 \times 2}{8.31 \times 293}$ ✓ $= 0.000222$ mol **X**

MARKER NOTE
Rhodri gains the 1st mark but not the 2nd because he has failed to notice the correct multipliers in the units on the graph.
1 mark

(ii) $T = \dfrac{pV}{nR} = \dfrac{1.2 \times 0.9}{0.000222 \times 8.31} = 585$ ✓

$T = \dfrac{pV}{nR} = \dfrac{0.27 \times 0.9}{0.000222 \times 8.31} = 132$ ✓

MARKER NOTE
Rhodri has applied the equation correctly for the 1st mark and has obtained the correct answer for the 3rd mark (notice how the powers of 10 slips cancel out in this section). Rhodri does not gain the 2nd mark because he has read the pressure scale incorrectly (the pressure should be 1.02×10^5 Pa).
2 marks

(iii) $U = \frac{3}{2} nRT = 1.5 \times 0.000222 \times 8.31 \times 585$ ✓

$= 1.08$ J **X**

MARKER NOTE
Here, Rhodri has not calculated the change in internal energy, only the internal energy at the (wrong) higher temperature. Nonetheless, he has used the equation and gains a generous mark.
1 mark

(iv) Counting squares, I guess there are 13.5 large squares below the curve ✓

Each square is $0.2 \times 0.2 = 0.04$ J

Work done = $0.04 \times 13.5 = 0.54$ J

MARKER NOTE
Rhodri's final answer is correct after he has multiplied his answer by a factor of 100. His explanation of why he has done this is given bod by the examiner. He has also stated that the work is done on the gas.
3 marks

But pressure is $\times 10^5$, volume $\times 10^{-3}$ so work done is 54 J. Also, this work is being done on the gas not by it. ✓✓ bod $\times 100$ (ecf)

(v) $\Delta U = 1.08$ J and W $= -54$ J

$\Delta U = Q - W$ so $Q = \Delta U + W = -53$ J ✓

So it seems to me that Teilo is no Einstein and he has got it wrong. ✓ ecf

MARKER NOTE
Rhodri's previous ΔU is wrong but he cannot be penalised for this again. He has found a good method of checking, i.e. applying the 1st law and his conclusion is correct with ecf.

2 marks

(vi) work done for CA $= 0.27 \times 1.1 = 0.297$ J ✓ bod

So work done $= -54 + 29.7 = -24.3$ J ✓

I suppose I'm meant to use $\Delta U = Q - W$ but I can't work out $\Delta U = Q - W$

MARKER NOTE
Rhodri's workings for the work done are correct. His numbers are not exactly the same as the MS but within tolerance. He cannot gain any more marks because he does not realise that $\Delta U = 0$ for a closed cycle.

2 marks

Total	11 marks /19

Ffion's answers

(a) ΔU is the increase in internal energy ✓

Q is heat entering the system ✓

W is work done by the gas ✓

MARKER NOTE
Ffion's answers are excellent, and she gains full marks. Note that 'change' is always defined as 'final − initial' so that change and increase are equivalent. Ffion is inconsistent in her use of 'system' and 'gas' but this is not penalised.

3 marks

(b) (i) $n = \dfrac{pV}{RT} = \dfrac{0.27 \times 10^5 \times 2.00 \times 10^{-3}}{8.31 \times 293}$ ✓

$= 0.0222$ mol ✓

MARKER NOTE
Ffion's answer is correct and she gains both marks.

2 marks

(ii) $\dfrac{p_A V_A}{T_A} = \dfrac{p_B V_B}{T_B}$, ✓ so

$T_B = \dfrac{p_B V_B}{p_A V_A} T_A = \dfrac{1.05 \times 0.90 \times 293}{0.27 \times 2.00} = 513$ K

$T_C = \dfrac{V_C}{V_A} T_A = \dfrac{0.90 \times 293}{2.00} = 132$ K

and $T_C = 132$ K ✓✓

MARKER NOTE
Ffion uses the $\dfrac{pV}{T}$ = constant form of the ideal gas law. She gains full credit for this. Note that she doesn't need to use the multipliers because they cancel when the ratios of the pressures and volumes are used in the calculations.

3 marks

(iii) $\Delta U = \frac{3}{2} nR (513 - 293) = 61$ J ✓✓

MARKER NOTE
Ffion's answer gains full marks even though it is wrong at first glance. This is where the examiner has to check Ffion's numbers from a previous answer and find that she has done everything correctly.

2 marks

(iv) Approximating AB to a straight line

Area $= 0.5 \times (1.02 + 0.27) \times 10^5 \times 1.1 \times 10^{-3}$

$= 71$ J ✓

which is close to 60 J   QED

MARKER NOTE
Ffion's approximation of the area is too large – she has approximated AB to a straight line. She has not explained the negative sign either and only gains the 1st mark for knowing that the area under the curve is the work.

1 mark

(v) in $\Delta U = Q - W$, we have

$61 = Q + 71$ So $Q = 10$ J ✓ bod

And the heat transferred is small in comparison so Teilo is quite accurate. ✓

MARKER NOTE
The examiner could have penalised Ffion because her previous W has changed sign. Bod was awarded because the examiner believes that Ffion is correcting her previous mistake with the sign.

2 marks

(vi) $\Delta U = 0$ because it's a closed cycle ✓

WD for CA $= 0.27 \times 10^5 \times 1.1 \times 10^{-3} = 30$ J ✓

So total WD $= 30 - 70 = 40$ J ✓ ecf

This is also the heat because $Q = W$ if internal energy doesn't change ✓

MARKER NOTE
Ffion's answer is perfect even though her final answer is a little large. This is just a knock-on effect from part (iv) and she deserves full marks with ecf.

4 marks

Total	17 marks /19

Component 2: Electricity and the Universe

Section 1: Conduction of electricity

Topic summary

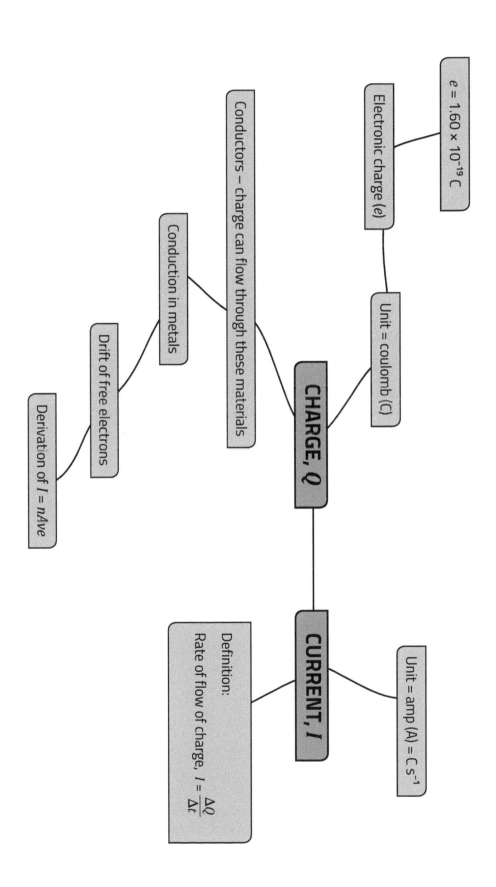

Q1 A light-emitting diode (LED) conducts a current of 15 mA. Calculate the number of electrons which flow into (or out of) the LED per minute. [$e = 1.60 \times 10^{-19}$ C] [2]

Q2 State what is meant by an electrical conductor. [1]

Q3 A capacitor is a device which stores electrical potential energy. It does this by holding equal positive and negative charges, Q, apart. The energy, W, stored is related to Q by the equation:

$W = \dfrac{Q^2}{2C}$ where C is a constant called the capacitance. The unit of C is the farad (F).

Express the farad in terms of the base SI units, m, kg, s and A. [4]

Q4 The radioactive material Am–241 is an alpha (α) emitter. A small sample of Am–241 has an activity of 37 kBq [that is, it emits 37×10^3 alpha particles per second]. Calculate the magnitude of the electric current that represented by this activity. [2]

Q5 Wire A is made from a metal with 3.0×10^{28} free electrons per m³. Its diameter is 0.60 mm and it carries an electric current of 1.5 mA. The values for wire B are 1.0×10^{28} m⁻³, 0.30 mm and 10 mA. Calculate the ratio v_A/v_B, where v_A and v_B are the drift velocities of the free electrons in wire A and wire B respectively. [3]

Q6 A light-dependent resistor (LDR) is an electronic device made from a material in which there are no free electrons in the dark. However, photons of light can temporarily eject electrons from atoms, so that they can move through the material.

Nigel and Iestyn decide to investigate the current, I, in an LDR when it is illuminated by a light source at different distances, d.

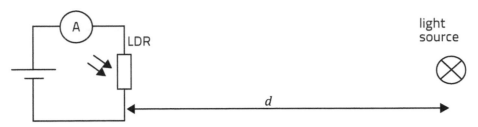

(a)   Nigel says that he expects the current to be inversely proportional to the square of the distance from the light source. Explain whether or not you agree with this statement.   [3]

....

....

....

....

....

(b)   The students' results are plotted below.

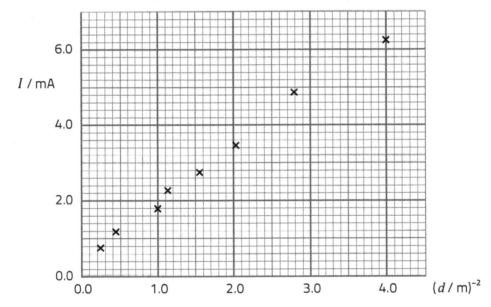

Evaluate to what extent the results agree with Nigel's hypothesis and suggest an explanation for any discrepancies.   [4]

....

....

....

....

....

....

Question and mock answer analysis

Q&A 1 Since May 2019, the charge on the proton has been defined as exactly $1.602\,176\,634 \times 10^{-19}$ C. The electronic charge is equal in magnitude but negative.

(a) A wire carries a current of (1.000 ± 0.005) mA. Evaluate what this means in terms of electron flow. [4]

(b) The current, I, in a metal wire is given by the equation:

$$I = nAve.$$

(i) Derive the equation. You may find it useful to include a diagram in your answer. [4]

(ii) Calculate the drift velocity of the electrons in an aluminium wire of diameter 0.50 mm which carries a current of 0.30 A, given that each atom of aluminium contributes three conduction electrons. [4]

Data: Density of aluminium = 2 710 kg m^{-3}; mean mass of an aluminium atom = 4.48×10^{-26} kg

What is being asked

There are not many questions that an examiner can ask which contain concepts only from within this very short topic. This one almost succeeds but it does require knowledge of density, which is from Topic 1.1, Basic Physics. Part (a) is a calculation based upon the relationship between current and charge flow, with the additional requirement to take the uncertainty of current into account. The requirement to bring together the current/charge equation, $Q = It$, and electronic flow rates makes this AO3. Part (b) starts with the standard derivation of $I = nAve$, which you are expected to know. This is followed by the application of $I = nAve$ to the case of an aluminium wire. The tricky aspects here are the correct application of SI multipliers and remembering to use the radius of the wire, rather than the diameter, in calculating the cross-sectional area. Combining the density and atomic mass to give the concentration of electrons is quite demanding.

Mark scheme

Question part		Description	AOs			Total	Skills	
			1	2	3		M	P
(a)		Current = charge flow per second [1] or by impl. $$\text{Electrons / s} = \frac{(1.000 \pm 0.005) \times 10^{-3}\ \text{A}}{1.602\ldots \times 10^{-19}\ \text{C}}\ [1]$$ $(= 6.242 \times 10^{15})$ Uncertainty = 0.5%, so electron flow $(6.24 \pm 0.03) \times 10^{15}\ \text{s}^{-1}$ [1] ecf With sf as above. [Accept $(6.242 \pm 0.031) \times 10^{15}$] [1]			4	4	3	
(b)	(i)	Drift distance in time $t = vt$ [1] Volume of wire with this length = Avt [1] Number of electrons drifting past any cross-section per second $[= nAvt /t] = nAv$ [1] $\therefore$ Current [= charge per second] = $nAve$ [1] Alternatively, considering a period of 1 second can score full marks.	4			4		
	(ii)	In 1 m^3 $n = 3$ [1] $\times \dfrac{2710\,\text{kg}}{4.48 \times 10^{-26}\,\text{kg}}$ [1] $(= 1.81 \times 10^{29})$ (or impl) $A = \pi \times (0.25 \times 10^{-3}\ \text{m})^2\ (= 1.96 \times 10^{-7}\ \text{m}^2)$ [1] (or impl) $v = 5.3 \times 10^{-5}\ \text{m s}^{-1}$ [1] ecf on A and n	1	2	1	4	4	
Total			**5**	**3**	**4**	**12**	**7**	

Rhodri's answers

(a) 1 A = 1 coulomb per second

so 1 mA = 10^{-3} C s^{-1} ✓

1 C = 1.602 176 634 × 10^{19} electrons

So 1 mA = 1.602 176... × 10^{16} electrons per second ✗

So 0.005 mA → 0.008 × 10^{16} electrons.

So there are between 1.610 and 1.594 × 10^{16} electrons
per second. ✓✓ ecf

MARKER NOTE

This answer is very good but Rhodri has made one mistake. He has obtained the first mark, equating current to rate of charge flow. His mistake is to say that 1/1.6 × 10^{-19} is 1.6 × 10^{19} electrons, so he lost a single mark (the second one). The other working is not as on the mark scheme, but he gives an equivalent answer: the range of electron flows consistent with the information and so he obtains the last two marks on the ecf principle.

3 marks

(b) (i) In a section of wire of length vt
the number of electrons is nAvt ✓

∴ Q = nAvte

I = $\dfrac{Q}{t}$ = $\dfrac{nAvet}{t}$ ✓ = nAve

MARKER NOTE

At first sight, Rhodri's answer looks good but he doesn't explain what he is doing. What is the significance of the distance vt, for example? He doesn't identify any of his symbols. A better approach for him would have been to draw and label a diagram, such as:

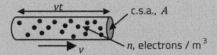

then use it to say
- all the electrons, numbering nAvt, within the volume will pass through the shaded area in time t.
- these electrons have charge, Q = nAvte
- and so the current, charge flow per second = nAve.

2 marks

(ii) I = nAve = 0.30 A

v = $\dfrac{I}{nAe}$

In 1 m³ the mass = 2710 kg

∴ n = $\dfrac{2710\,kg}{4.48 \times 10^{-26}\,kg}$ = 6.05 × 10^{28} ✗ ✓

A = π × (0.0025)2 = 1.96 × 10^{-5} ✗

So v = $\dfrac{0.30}{6.05 \times 10^{28} \times 1.96 \times 10^{-5} \times 1.60 \times 10^{-19}}$

= 1.58 × 10^{-6} m s^{-1} ✓ecf

MARKER NOTE

Rhodri correctly works out the number of aluminium atoms per m³ (the first ✓) but apparently identifies this with n, so doesn't multiply by 3 (hence the first ✗). His method for calculating A is correct but unfortunately, he doesn't convert from mm to m correctly and he loses the A mark (the second ✗). However, the mark for calculating the drift velocity is available and he is awarded it on the ecf principle.

2 marks

Total	7 marks /12

Ffion's answers

(a) Current = charge per second. ✓

So $1.000 \text{ mA} = 1 \times 10^{-3} \text{ C s}^{-1}$

$1 \text{ C} = \dfrac{1}{1.602\,176\,634 \times 10^{-19}}$ electrons

$= 6.241\,509\,074 \times 10^{18}$ e/s ✓

$\therefore 1 \text{ mA} = 6.241\,509\,074 \times 10^{15}$ e/s

If 1.005 mA, then 6.274×10^{15}

So Ans $= 6.242 \pm 0.03 \times 10^{15}$ e / s ✓ ✗

> **MARKER NOTE**
>
> Ffion's answer gains the first three marks. The first two are as in the mark scheme. Her method of determining the uncertainty in the number of electrons per second is to find the maximum flow rate, i.e. with a current of 1.005 mA. The uncertainty is then this maximum minus the value with $I = 1.000$ mA. To gain the last mark, she should have expressed the flow rate to the same decimal place as the uncertainty, as in the mark scheme.
>
> **3 marks**

(b) (i) n = number of electrons per m³

If the drift velocity is v all the electrons in a length vt will go past in time t. ✓

Volume of this length = Avt, ✓

so number of electrons = $nAvt$

So the number of electrons per second is nAv ✓

So the charge per second, which is the current, is $nAve$ ✓

> **MARKER NOTE**
>
> Ffion is able to communicate the logic of the derivation without using a diagram. She clearly identifies the meanings of n and v in the equation, and logically develops the sequence of ideas:
>
> - the volume of the significant length of wire, vt.
> - number of electrons passing a cross-section in a time, t
> - the charge on these electrons
> - hence the expression for the current.
>
> **4 marks**

(ii) Atoms/m³ $= \dfrac{2710 \text{ kg/m}^3}{4.48 \times 10^{-26} \text{ kg/atom}}$

$= 6.049 \times 10^{28}$ ✓

3 electrons per atom,

so $n = 1.815 \times 10^{29}$ per m³ ✓

$A = \pi \times \left(\dfrac{5 \times 10^{-4}}{2}\right)^2 = 1.96 \times 10^{-7} \text{ m}^2$ ✓

$\therefore v = \dfrac{I}{nAe} = 1.76 \times 10^{-5} \text{ m s}^{-1}$ ✗

> **MARKER NOTE**
>
> An almost perfect answer from Ffion but with an unfortunate slip at the end. She sets out her working clearly, calculating first the number density of the aluminium atoms (first mark) and then multiplying it by 3 (second mark) to give n. She correctly applies $A = \pi r^2$ for the third mark.
>
> At the end she uses the incorrect value n, which appears to be a mistake but costs her a mark.
>
> **3 marks**

| Total | 10 marks /12 |

Section 2: Resistance

Topic summary

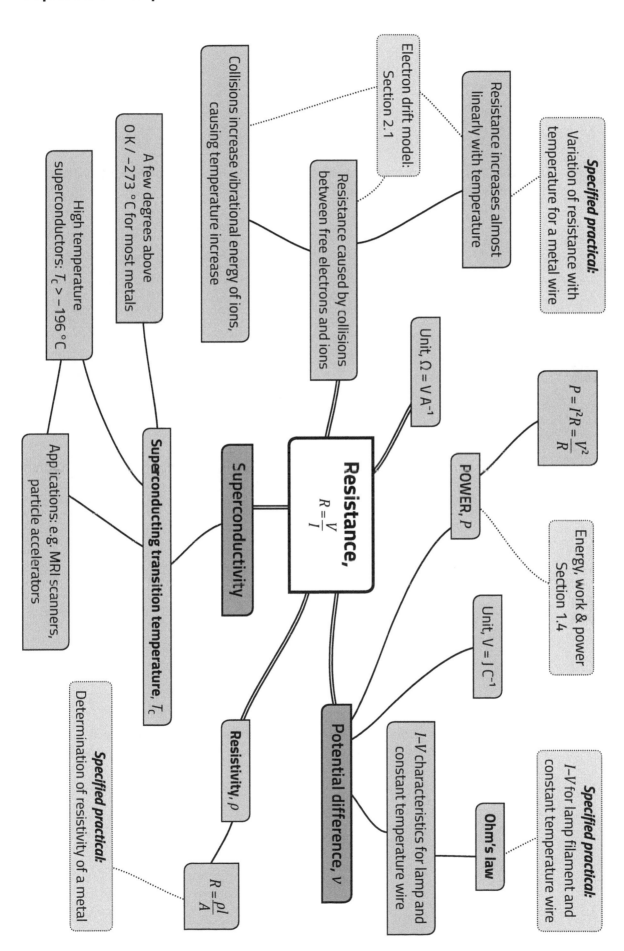

Component 2 Practice questions

Resistance,
$$R = \frac{V}{I}$$

Unit, $\Omega = V\,A^{-1}$

Specified practical:
Variation of resistance with temperature for a metal wire

Resistance increases almost linearly with temperature

Electron drift model:
Section 2.1

Collisions increase vibrational energy of ions, causing temperature increase

Resistance caused by collisions between free electrons and ions

A few degrees above 0 K / −273 °C for most metals

High temperature superconductors: $T_C > -196\,°C$

Applications: e.g. MRI scanners, particle accelerators

Superconducting transition temperature, T_C

Superconductivity

Resistivity, ρ

Specified practical:
Determination of resistivity of a metal

$$R = \frac{\rho l}{A}$$

Potential difference, V

Unit, $V = J\,C^{-1}$

I-V characteristics for lamp and constant temperature wire

Ohm's law

Specified practical:
I-V for lamp filament and constant temperature wire

POWER, P

$$P = I^2 R = \frac{V^2}{R}$$

Energy, work & power
Section 1.4

Q1 A potential difference (pd) is applied across a component, X. As a result, there is a current of 1.5 A in X and 300 J of energy is transferred in a time 20 s.

Starting from definitions of potential difference and current, deduce the pd across X with clear reasoning. [3]

....

....

....

....

Q2 The joule (J), the SI unit of energy, can be expressed as kg m^2 s^{-2}. Starting from definitions of pd and charge, express the volt (V) in terms of SI base units. [3]

....

....

....

....

Q3 (a) State Ohm's law. [1]

....

....

(b) Liam and Paul disagree over whether the equation $V = IR$ is a statement of Ohm's law. Explain to what extent it is. [2]

....

....

....

Q4 The current, I, in a wire can be related to the drift of free electrons by the equation $I = nAve$.

(a) Identify the symbols n, A, e and v in the equation. [2]

....

....

(b) A pd is applied across a wire and a current produced. If the temperature of the wire increases, the resistance of the wire increases. Account for this in terms of the electron drift model. [4]

....

....

....

....

....

Q5 Many metals exhibit superconductivity.

(a) State briefly what is meant by superconductivity. Include a graph and label the axes and significant features. [3]

....

....

....

....

....

(b) State what is meant by high-temperature superconductors and explain their advantage in a named application. [3]

....

....

....

....

....

....

Q6 A car filament lamp is designed to operate at 12 V. When the pd across it is 2.4 V, the current in the lamp is 1.5 A. When it is operating at the rated pd, the current is 3.0 A.

(a) Calculate the following ratios:

(i) $\dfrac{\text{Resistance of the filament at 12 V}}{\text{Resistance of the filament at 2.4 V}}$ [2]

....

....

....

(ii) $\dfrac{\text{Power of the lamp at 12 V}}{\text{Power of the lamp at 2.4 V}}$ [2]

....

....

....

(b) Explain briefly, in terms of a model of conduction, why the temperature of the filament increases with pd. [2]

....

....

....

Q7 A reel of thin enamelled copper wire has a label with the following data:

0.1 mm   50 g   2.18 Ω/m   14 306 m/kg

(a) Select data from above and use it to estimate the resistivity of copper. [The 0.1 mm is the diameter.]

[3]

(b) A website gives the density of copper as 8.89 g cm⁻³. Evaluate whether this is in agreement with the above data.

[3]

Q8 Peter and Sion use a digital micrometer, a resistance meter and metre rule to determine the resistivity of constantan in the form of a wire. They obtained the following values:

diameter 0.32 ± 0.01 mm   length 2.000 ± 0.002 m   resistance 13.9 ± 0.1 Ω

(a) Use the data to determine a value for the resistivity along with its absolute uncertainty. Give your answer to an appropriate number of significant figures.

[4]

(b) They had another reel of constantan wire with a diameter of approximately twice that of their original wire. Peter said that they would get a value of resistivity with a lower uncertainty if they used this second reel because the percentage uncertainty in the diameter would be less. Sion said they'd be better just using a longer piece of wire.

Evaluate their suggestions.

[4]

Q9 A student finds an old filament-lamp labelled 240 V, 60 W. She uses a resistance meter to determine its room-temperature resistance. The result is 80 Ω. Estimate the temperature of the filament of the lamp in normal operation, assuming that the resistance of the filament is approximately proportional to its absolute temperature. [3]

....

....

....

....

....

Q10 The owner of a tropical fish tank wants to make a 10 W, 30 V electrical heater to maintain the water temperature. He decides to use a heater wire made from constantan, an alloy with an almost constant resistance over a wide range of temperatures.

(a) Explain an advantage of using a material with a constant resistance when making a heater filament. [2]

....

....

....

(b) The constantan wire has a diameter of 0.12 mm. Calculate the length of wire which the owner should use for the heater.
Resistivity of constantan = 4.9×10^{-7} Ω m. [4]

....

....

....

....

....

....

Q11 Wire **A** has diameter D, length l and is made from a material of resistivity ρ.

Wire **B** has a diameter $2D$, length $3l$ and is made from a material of resistivity 2.5ρ.

The two wires are separately connected to power supplies with the same pd, V.

Giving your reasoning, determine the ratio

$$\frac{\text{power dissipated in wire } \mathbf{A}}{\text{power dissipated in wire } \mathbf{B}}$$ [3]

....

....

....

....

Component 2 Practice questions

Q12 (a) A coil of iron wire is contained in a test tube of oil with its ends projecting. Describe briefly a method for investigating the variation of the resistance of this wire over a range of temperatures between 0 °C and 100 °C. Sketch a graph of the expected results. [4]

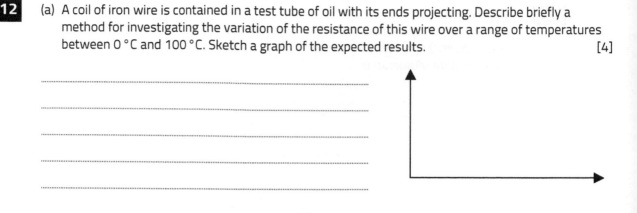

(b) A coil of iron wire has a resistance of 12.0 Ω at 20 °C. At 75 °C the resistance is 16.3 Ω. When the wire is placed in a bath of hot engine oil, its resistance is 18.7 Ω. Estimate the temperature of the oil, stating any assumption you make. [3]

Q13 *Point contact diodes* were developed in the 1940s for use in microwave receivers for radar. A website gives the following sketch graph of the variation of resistance with applied pd for a point contact diode.

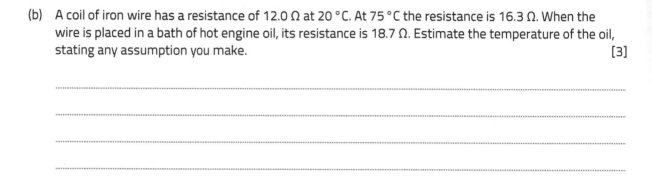

Sketch a current–voltage characteristic for a point contact diode, showing the features relating to V_1 and V_2. [3]

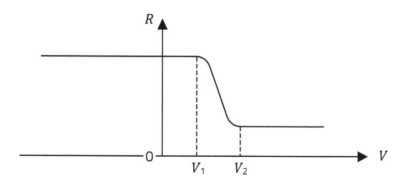

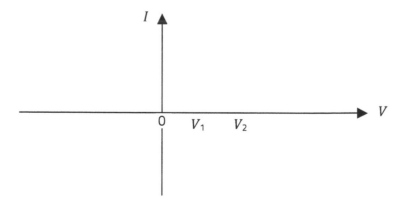

Question and mock answer analysis

Q&A 1 A group of students investigates the current–pd (*I–V*) graph of an old 12 V filament car headlamp. They obtain the following results.

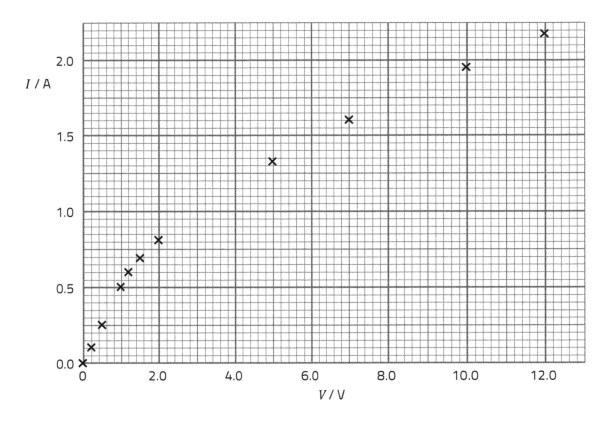

(a) With the aid of a circuit diagram, describe how they could have obtained these results. [3]

(b) From their results, describe how the **resistance** of the filament changes with pd. Include calculations in your answer. [4]

(c) Explain qualitatively, in terms of electrons, the variation of resistance with pd between 2 V and 12 V. [4]

(d) Use the graph to determine the pd at which this light bulb filament would dissipate the same power as a 4.0 Ω resistor and state the value of this power. [3]

What is being asked

Although the question is based upon a specified practical, only part (a) directly examines this, with a standard AO1 recall description. Part (b) looks like a standard piece of graph description. However there are two points to note, which make this an AO3 question: first, the variation of <u>resistance</u> is asked for, not current (i.e. the variable on the *y* axis); second, the graph has a straight line part and a curve. Part (c) is a standard piece of bookwork, albeit not easy, hence AO1. The comparison of the lamp with an ohmic resistor, (d), is a fairly straightforward part of evaluation (AO3).

Mark scheme

Question part	Description	AOs			Total	Skills	
		1	2	3		M	P
(a)	Circuit drawn with lamp connected across a power supply [1]						
	Method of adjusting pd / current, e.g. rheostat or variable voltage supply [1]	3			3		3
	Ammeter in series and voltmeter in parallel with the lamp [1]						
(b)	Between 0 and 1.2 V (or 0 and 0.6 A) the resistance is constant [1] at 2.0 Ω [1]						
	At higher voltages (or currents) the resistance increases [steadily] [1] with identified value at specified voltage (or current), e.g. 5.5 Ω at 12 V [1]			4	4	2	
	NB 3 max If no explicit calculation, e.g. $\frac{12\,V}{2.17\,A} = 5.5\,Ω$						
(c)	Resistance is caused by collisions between free (or conduction) electrons and metal atoms / ions / lattice [1]						
	At higher currents, more energy is passed on in the collisions, raising the temperature [1]	4			4		
	At higher temperature the time between collisions is less [1] so the drift velocity is lower and the current is therefore lower [than if the temperature were constant]. [1]						
(d)	Graph line drawn for (appropriate section of) the results and I–V graph for a 4 Ω resistor drawn (passing through 4.0, 1.0) [1] or equiv.						
	Intersection of graphs identified – 5.8 V [1]			3	3	1	
	Power = 8.4 W ecf [1]						
Total		**7**	**0**	**7**	**14**	**3**	**3**

Rhodri's answers

(a)

✓✓

- Set up circuit
- Adjust voltage to zero – measure current
- Increase voltage in stages and read current each time. ✓

MARKER NOTE

A good answer which hits all the marking points

3 marks

(b) The resistance increases as the voltage increases. ✓

e.g. at 2 V,

$R = \frac{2}{0.8} = 2.5\,Ω$

At 12 V $R = \frac{12}{2.15} = 5.6\,Ω$ ✓

MARKER NOTE

Rhodri has missed the fact that the low voltage points lie on a straight line, giving a constant resistance. Hence he cannot gain the first two marks. He gains the other points because he correctly states the variation of resistance and calculates the value at 12 V.

2 marks

(c) At higher voltages and currents, the electrons are moving faster so the wire is at a higher temperature. ✓ The resistance of a metal wire increases with temperature, so the higher the voltage the greater the resistance.

(d)

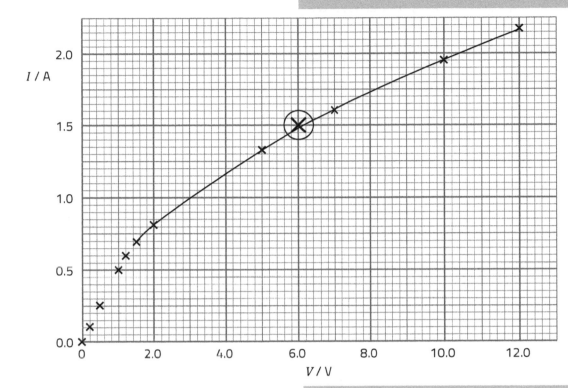

See graph for lamp

The bulb needs to have a resistance of 4 Ω – so 6 V, 1.5 A does this. ✗ [not enough]

$$Power = \frac{V^2}{R} = \frac{6^2}{4} = 9 \text{ W} ✓ \text{ ecf}$$

Total **7 marks /14**

Ffion's answers

(a)

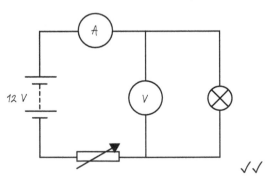

In this circuit, set the variable resistor to its maximum resistance and note the current and voltage readings.

Adjust the variable resistor to a series of lower values and note the voltage and current readings. ✓

Component 2 Practice questions

(b) Up to about 1.0 V, the graph is a straight line so the resistance is constant. ✓ Above 1.0 V, the graph curves to become more nearly horizontal so the resistance goes up with voltage. ✓

At 12 V the resistance is 5.5 Ω ✓ X

(c) At higher voltages, the electrons gain more kinetic energy between collisions with metal ions, so they pass on more energy, increasing the temperature. ✓ At higher temperatures because the random speed of the electrons is greater the time between collisions is less. ✓ Hence the mean drift speed is less and the current is less (than if the temperature was lower) ✓ so the resistance is higher. ✓

(d)

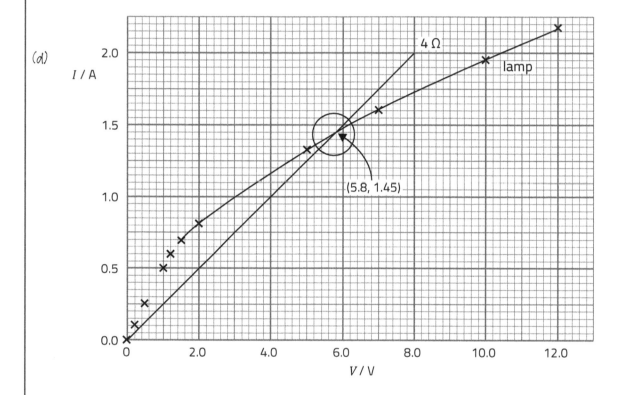

These are the I–V graphs for the lamp and a 4 Ω resistor. ✓

The two dissipate the same power where V and I are the same, i.e. when they cross (because P = VI). So the voltage is 5.8 V. ✓

So power = VI = 5.8 V × 1.45 A = 8.4 W ✓

Section 3: DC Circuits

Topic summary

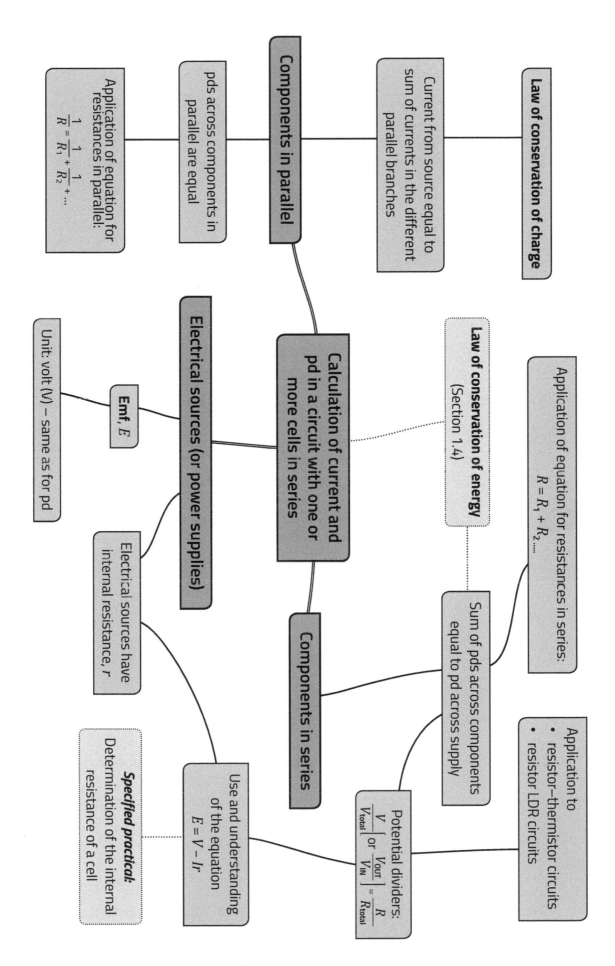

Components in parallel

Application of equation for resistances in parallel:
$$\frac{1}{R} = \frac{1}{R_1} + \frac{1}{R_2} + \dots$$

pds across components in parallel are equal

Current from source equal to sum of currents in the different parallel branches

Law of conservation of charge

Calculation of current and pd in a circuit with one or more cells in series

Electrical sources (or power supplies)

Unit: volt (V) – same as for pd

Emf, E

Electrical sources have internal resistance, r

Law of conservation of energy (Section 1.4)

Application of equation for resistances in series:
$$R = R_1 + R_2 \dots$$

Sum of pds across components equal to pd across supply

Components in series

Application to
• resistor–thermistor circuits
• resistor LDR circuits

Potential dividers:
$$\frac{V}{V_{total}} \ \text{or} \ \frac{V_{OUT}}{V_{IN}} = \frac{R}{R_{total}}$$

Use and understanding of the equation
$$E = V - Ir$$

Specified practical: Determination of the internal resistance of a cell

Q1 In the circuit, the lamps $L_1 - L_3$ are identical. The greater the current, the brighter is the lamp.

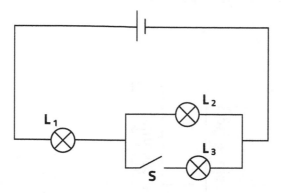

(a) With the switch, **S**, open as shown, explain why lamps L_1 and L_2 are equally bright. [2]

....

....

....

(b) Explain the changes in brightness of the three lamps when **S** is closed. [4]

....

....

....

....

....

....

....

Q2 In the circuit, the values of pd shown were obtained by connecting a voltmeter across each resistor. The battery has negligible internal resistance.

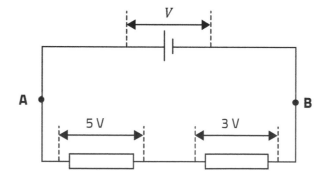

(a) Use the principle of conservation of energy to explain why the pd across the battery is 8 V. [2]

....

....

....

(b) A third resistor is connected directly between **A** and **B**. State the pd across it. .... [1]

Q3   You are provided with three 12 Ω resistors. What different values of current can you take from a 6.0 V power supply using different combinations of some or all of these resistors? You should sketch the combinations in the space below.   [4]

Q4   A circuit is connected as shown. The power supply has negligible internal resistance.

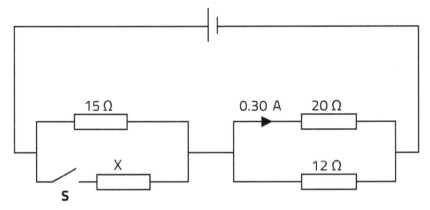

(a)   Calculate the power dissipated in the circuit when the switch, **S**, is open as shown.   [4]

....

....

....

....

....

(b)   Without calculation, explain how the pd across the 15 Ω resistor changes when switch **S** is closed.   [3]

....

....

....

....

....

Q5 The circuit shows the sensing circuit of a light alarm. The terminals labelled V_{OUT} are connected to the alarm input, which has a very high resistance.

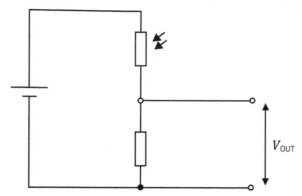

(a) Explain how the output voltage, V_{OUT}, varies with light level incident upon the LDR. [4]

....

....

....

....

....

....

(b) The resistance–temperature graph for a thermistor is shown.

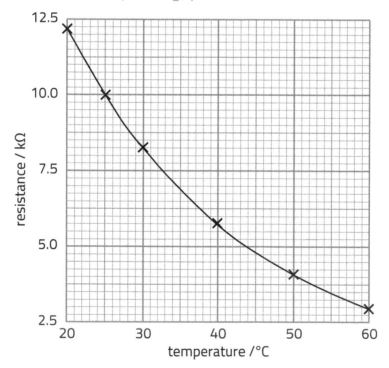

In the space to the right of the graph, **draw a sensing circuit** for a low-temperature alarm which should operate when the temperature drops below 37 °C. The power supply has a terminal pd of 12.0 V and the alarm is triggered when V_{OUT} from the sensor rises above 5.0 V. [4]

Q6 The emf of an electric battery is 9.0 V.

(a) Explain what is meant by an emf of 9.0 V. [2]

(b) The battery produces a current of 1.5 A when it is connected into a circuit. The pd across the battery terminals is then 7.8 V.

(i) Account for the energy transfers in the battery and external circuit. [3]

(ii) Determine the internal resistance of the battery. [1]

Q7 A power supply has a terminal pd of 6.5 V when a 10 Ω resistor is connected across it. When a second 10 Ω resistor is connected in parallel with the first, the pd falls to 6.0 V. Determine the emf and internal resistance of the power supply. [4]

Q8 It can be shown that an electrical power supply transfers the maximum power to a circuit when the external resistance, R, is equal to the internal resistance.

A conventional (non-rechargeable) cell has an emf of 1.5 V and an internal resistance of 0.3 Ω. A rechargeable Ni–Cd cell has an emf of 1.2 V and an internal resistance of 35 mΩ. Compare the maximum power available from these cells. [4]

Q9 In a practical lesson, a student is given a sealed power supply consisting of a battery of unknown emf and resistor of unknown value, r, in series. They investigate the power supply using this circuit.

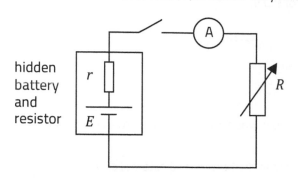

The variable resistance, R, is made by using some or all of the following resistors:

3.3 Ω, 10 Ω, 18 Ω

They obtained the following results:

R / Ω	3.3	6.4	7.6	10.0	13.3	18.0
I / A	0.43	0.34	0.31	0.27	0.23	0.19
$(I / A)^{-1}$						

(a) Show how they used the available resistors to obtain the 7.6 Ω resistance. [2]

....

....

....

(b) Starting from the equations:

$$V = E - Ir \qquad \text{and} \qquad V = IR,$$

where V is the pd across the terminals of the power supply, show that a graph of $1/I$ against R should be a straight line of gradient $1 / E$. [3]

....

....

....

(c) Complete the table by adding in a row of values of I^{-1}, to an appropriate number of significant figures. [2]

(d) Use the grid on the next page to plot a graph of I^{-1} (on the y axis) against R. [4]

(e) The teacher tells the students that the emf of the battery is 4.8 V and the internal resistor has a value of 8.2 Ω. Evaluate whether this is consistent with the students' results. [4]

....

....

....

....

....

....

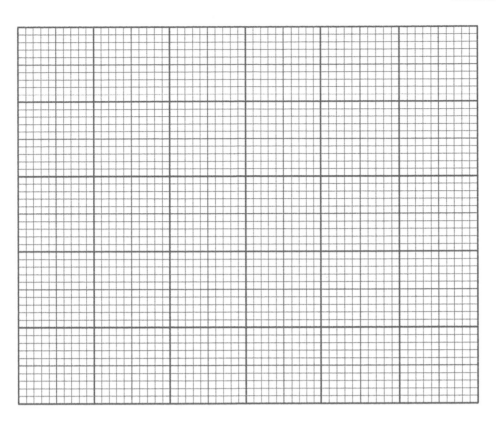

Q10 A 2.5 V, 1.5 W filament lamp is powered using 3 Ni-Cd cells, each of emf 1.2 V and negligible internal resistance, connected in series together with a resistor as shown in the circuit.

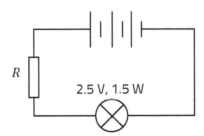

(a) Calculate the resistance, R, required for the lamp to be powered at its rated value. [3]

....

....

....

....

....

(b) Determine the fraction of the output power of the battery which is transferred within the resistor. [2]

....

....

....

Q11 The silicon diode is a semiconductor device which, in normal operation, conducts in one direction only – shown in the following diagram of its symbol.

An approximate *I–V* characteristic for a diode is shown. Notice that, at the 'turn-on voltage' of 0.7 V, any value of current is possible.

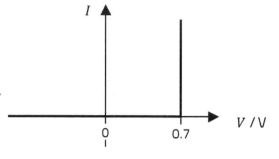

(a) A silicon diode is connected in series with a resistor of value 820 Ω. The two carry a current of 30 mA. Calculate the pd across the ends of the two. A diagram might help your answer. [2]

....

....

....

(b) A red light-emitting diode (LED) has a similar characteristic to the silicon diode. Its turn-on voltage is 1.9 V. The LED is to be used as an On-indicator for an electrical machine. The indicator circuit has a voltage supply of 9.0 V. The current in the LED must be between 10 mA and 25 mA in operation. Determine which of the following resistors are suitable for use as the series resistor. Give your reasoning.

68 Ω   100 Ω   220 Ω   470 Ω   680 Ω   1.0 kΩ   1.5 kΩ [3]

....

....

....

....

....

Question and mock answer analysis

Q&A 1   A battery has an emf of 9.0 V and internal resistance 1.8 Ω.

(a)   Explain what is meant by 'an emf of 9.0 V'.   [2]

(b)   The battery is connected to a wire of resistance 5.4 Ω. Determine:

   (i)   The rate at which chemical energy is being transferred in the cell.   [2]

   (ii)   The fraction of this energy which is transferred to the wire.   [2]

(c)   Sketch a graph of the terminal pd of this battery against the current delivered. Label the graph to show how particular features relate to the values of emf and internal resistance.   [4]

What is being asked

This is quite a short question and seeks to determine whether you understand the concepts of emf and internal resistance. In part (a), the examiner could have asked, 'Explain what is meant by emf.' This would have been an AO1 question and required you to write the textbook definition. Asking it in this way requires you to use the data in an appropriate way and is therefore AO2. Part (b) has two calculations. which build on the answer to part (a). Again, it is AO2 because you need to apply your knowledge to a particular situation. Part (c) is a straightforward piece of bookwork, again applied to this battery. There are quite a few marks available, so it is clear that the examiner expects rather a lot of detail.

Mark scheme

Question part			Description	AOs			Total	Skills	
				1	2	3		M	P
(a)			9.0 J of energy are transferred from chemical to electrical [1] … per coulomb [of charge] {entering / passing through/ leaving} the battery [1]		2		2		
(b)	(i)		Current $\left[= \dfrac{9.0\,V}{5.4\,\Omega + 1.8\,\Omega} \right] = 1.25\,A$ [1] Rate of transfer [= 9.0 V × 1.25 A] = 11.25 W [1]		2		2	2	
	(ii)		Power dissipated in wire = I^2R = 8.43 W [1] ecf Fraction $\left[= \dfrac{8.43\,W}{11.25\,W} \right] = 0.75$ [1] **Alternative answer** [Using potential divider] Fraction of total $V = \left[= \dfrac{5.4\,\Omega}{5.4\,\Omega + 1.8\,\Omega} \right] = 0.75\,(\checkmark)$ I same in R and r so fraction of power = 0.75 $(\checkmark)$		2		2	2	
(c)			Axes drawn and labelled, (V/V and I/A), straight-line graph sloping down from a +V intercept [1] V intercept labelled 9[.0] (accept E) [1] I intercept labelled 5[.0] (accept E / r) [1] [Accept statement that the maximum current is 5 or E / r] Gradient labelled -1.8 (accept -r) [1] Maximum mark for no correct number = 2 Maximum mark with only 1 correct number = 3	2	2		4	2	
Total				**2**	**8**	**0**	**10**	**6**	

Rhodri's answers

(a) The emf is the chemical energy lost per coulomb delivered by the battery ✗✓

MARKER NOTE

Rhodri's answer is a reasonable stab at the textbook definition of emf and gains the qualitative mark. He misses the mark which ties it in to the energy figure given.

1 mark

(b)(i) The current = emf/tot resistance

= 9.0 / 7.2

= 1.25 A ✓

Power in resistor = $I^2R = 1.25^2 × 5.4$

= 8.4 W

Power in cell = $I^2r = 1.25^2 × 1.8$

= 2.8 W

So total power = 11.2 W ✓

MARKER NOTE

Rhodri has not found the easiest way of answering this question but still gains both marks. He appears to realise that rate of energy transfer is power. After calculating the current, it would be much easier to use EI.

2 marks

(ii) The power in the wire = $\frac{V^2}{R} = \frac{9.0^2}{5.4}$ ✗

= 15 W

So % = $\frac{15}{11.2}$ $\frac{11.2}{15}$ × 100 = 74.6% ✗

MARKER NOTE

Rhodri has appreciated that the wire is in series with the internal resistance. However, he imagines that the pd across the wire is the emf, so loses the first mark. His power in the wire is greater than the total power transfer, so he cannot rescue the calculation.

0 marks

(c)

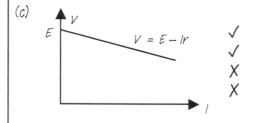

V = E − Ir   ✓
                ✓
                ✗
                ✗

MARKER NOTE

Again, Rhodri hasn't spotted the need to use the data given and give the graph for *this* battery, so only achieves the first two marks.

2 marks

Total   **5 marks /10**

Ffion's answers

(a) This means that for every coulomb of charge which passes through the battery ✓, 9.0 J of chemical energy is used to push the charge around the circuit. ✓ [bod]

MARKER NOTE

Both marks are given. The examiner is not impressed by the 'energy used' but has given it the benefit of the doubt.

2 marks

(b)(i) $I = \frac{E}{R+r} = \frac{9.0}{5.4+1.8} = 1.25$ A ✓

Rate = energy per second = EI

= 9×1.25 = 11.25 W ✓

MARKER NOTE

An exemplary answer from Ffion, who realises that EI is the rate of energy transfer in the battery.

2 marks

(ii) pd across wire = IR = 1.25 × 5.4 = 6.75 V

So power in wire = IV = 1.25 × 6.75

= 8.4375 W ✓

$\frac{8.4375}{11.25}$ × 100% = 75% ✓

MARKER NOTE

Ffion has, unnecessarily, calculated the pd across the wire. She could have used this directly as follows:

$$\frac{\text{power in wire}}{\text{total power}} = \frac{V_{\text{wire}}}{E} = \frac{6.75}{9.0} = 0.75$$

But she still picks up both marks.

2 marks

(c)

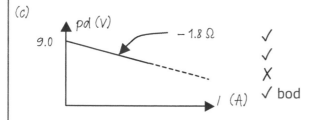

✓
✓
✗
✓ bod

MARKER NOTE

The only mark Ffion has not gained is the most difficult one – the intercept on the current axis. The 'bod' on the last mark is because the label on the graph doesn't clearly mention the gradient.

3 marks

Total   **9 marks /10**

Q&A 2 Sue and Ianto connect a resistance wire, **AB**, of length 75 cm and diameter 0.50 mm, into a circuit with a cell of emf 6.0 V and negligible internal resistance, as shown. They connect a voltmeter between **A** and a moveable point **P**.

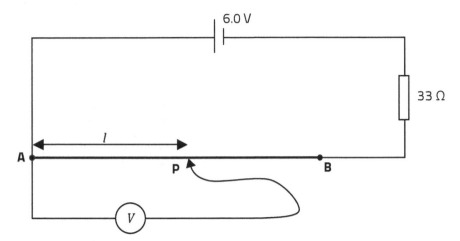

(a) Sue says that the voltmeter reading, V, should be proportional to the length, l, between **A** and **P**. Explain why she is correct. [Assume the voltmeter has infinite resistance.] [3]

(b) Ianto obtains a set of results of V and l and plots a graph, which has a gradient of 0.020 V cm^{-1}. Use this value to determine the resistivity of the metal of the wire. [5]

What is being asked

Part (a) of the question is a little unusual but it appears straightforward. You need to relate the variation of AP to potential divider ideas. It is not a standard explanation, hence it is AO3. In part (b) you can use the result of part (a), which is why the examiner asked you to show that Sue was correct, rather than whether she was correct. It then becomes a standard, if rather complex, AO2 calculation.

Mark scheme

| Question part | Description | AOs | | | Total | Skills | |
		1	2	3		M	P
(a)	There is no current through the voltmeter so the current is the same throughout AB [1] The resistance of AP is proportional to the length [1] Use of $V = IR$ or potential divider argument to show $V \propto l$ [1]			3	3		
(b)	$V_{AB} = 1.50$ V [1] pd across 33 Ω resistor = 4.5 V [1] Resistance of AB = 11.0 Ω [1] no ecf Rearrangement of $R = \dfrac{\rho l}{A}$ with ρ the subject at any stage [1] $\rho = 2.9 \times 10^{-6}$ Ω m ((**unit**)) [1] ecf on R_{AB}		5		5	4	
Total		0	5	3	8	4	

Rhodri's answers

(a) $R = \frac{\rho l}{A}$, so the resistance of the wire AP is proportional to the length ✓

So the voltage is also proportional to the length. ✗ not enough

MARKER NOTE

Rhodri has understood the importance of the current being the same at all points in the wire, so he can be awarded the first mark. To gain the last mark he needs clearly to tie in the voltage and resistance.

1 mark

(b) Current $= \frac{6.0}{33} = 0.182$ A ✗

$V_{AB} = 75 \times 0.020 = 1.5$ V ✓

Resistance of wire AB $= \frac{1.5}{0.182}$

$= 8.25 \ \Omega$ ✗

$\rho = \frac{RA}{l}$ ✓ ecf$= \frac{8.25 \times \pi (0.25 \times 10^{-3})^2}{0.75}$

So $\rho = 2.2 \times 10^{-6} \ \Omega \ m^{-1}$ ✗ unit

MARKER NOTE

Rhodri's first statement is incorrect. The 6 V pd is across the series combination of the wire and resistor. The pd across the 33 Ω resistor is 6.0 − 1.5 = 4.5 V. There is no ecf available for the resistance calculation. He rearranges the resistivity equation and obtains the mark but loses the last mark because of the incorrect unit.

2 marks

Total　　　　　　　　　　　　　　**3 marks /8**

Ffion's answers

(a) The current is constant so the pd across the wire is constant ✗ [not clear]. If AP is half AB then the resistance of AP is half the resistance of AB ✓. The wire acts as a potential divider, so the pd across AP would be half that across AB. ✓ Similarly if AP is $\frac{1}{4}$ AB, the pd will be $\frac{1}{4}$ that across AB, so the pd is proportional to the length of AB.

MARKER NOTE

This is almost a perfect answer but the examiner doesn't feel able to award the first mark. What does 'constant' mean in this answer? Does it mean constant in time or constant in position? However, the other two marks are awarded.

2 marks

(b) pd across 1 cm of wire = 0.02 V

∴ pd across AB $= 75 \times 0.020 = 1.50$ V ✓

∴ pd across 33 W $= 6.0 - 1.5 = 4.5$ V ✓

∴ Current $= \frac{4.5}{33} = 0.136$ A

∴ $R_{AB} = \frac{1.5 \ V}{0.136 \ A} = 11 \ \Omega$ ✓

$R = \frac{\rho l}{A}, \ \rho = \frac{11 \times \pi (0.0005)^2}{0.75}$ ✓ [rearrange]

∴ $\rho = 1.2 \times 10^{-5} \ \Omega \ m$ ✗

MARKER NOTE

This is a well set-out answer. Ffion calculates the resistance of AB using the calculated current. She could also have used a potential divider argument. They are equally valid.

Her only mistake is to use the diameter of the wire instead of the radius, so she obtains the rearrangement mark but not the mark for the final answer.

4 marks

Total　　　　　　　　　　　　　　**6 marks /8**

Section 4: Capacitance

Topic summary

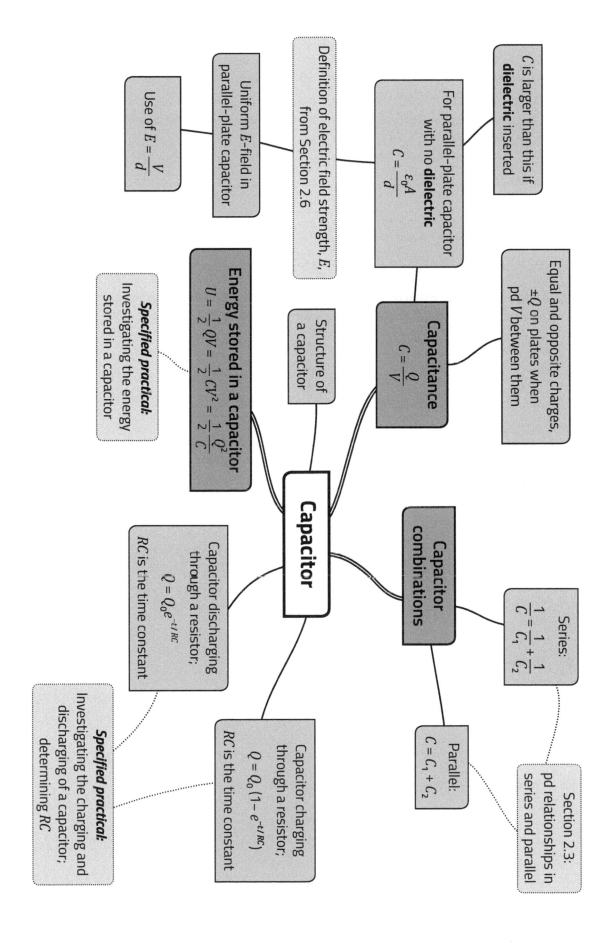

Capacitor

Use of $E = \dfrac{V}{d}$

Uniform E-field in parallel-plate capacitor

Definition of electric field strength, E, from Section 2.6

For parallel-plate capacitor with no **dielectric**

$$C = \frac{\varepsilon_0 A}{d}$$

C is larger than this if **dielectric** inserted

Equal and opposite charges, $\pm Q$ on plates when pd V between them

Capacitance

$$C = \frac{Q}{V}$$

Structure of a capacitor

Energy stored in a capacitor

$$U = \frac{1}{2}QV = \frac{1}{2}CV^2 = \frac{1}{2}\frac{Q^2}{C}$$

Specified practical: Investigating the energy stored in a capacitor

Capacitor discharging through a resistor;

$$Q = Q_0 e^{-t/RC}$$

RC is the time constant

Capacitor charging through a resistor;

$$Q = Q_0(1 - e^{-t/RC})$$

RC is the time constant

Specified practical: Investigating the charging and discharging of a capacitor; determining RC

Capacitor combinations

Series:
$$\frac{1}{C} = \frac{1}{C_1} + \frac{1}{C_2}$$

Parallel:
$$C = C_1 + C_2$$

Section 2.3: pd relationships in series and parallel

Q1 Determine the charges on each plate of a 22 mF capacitor when a pd of 12 V is placed between them. [2]

Q2 A parallel-plate capacitor of capacitance 500 pF is to be made using two flat metal plates, each measuring 10 cm × 10 cm.

(a) Calculate the required separation of the plates, if there is only air between them. [3]

(b) Ludovic intends to make the capacitor, but with the gap between the plates filled with an insulating polymer. Discuss whether the plate separation should be that calculated in (a). [2]

Q3 A pd of 30 V is applied between the plates of an air-spaced parallel-plate capacitor. Each plate has an area of 64 cm², and the plate separation is 0.40 mm. Calculate:

(a) The charge on either plate. [3]

(b) The energy stored in the capacitor. [2]

(c) The electric field strength in the gap between the plates. [1]

Q4 A battery is connected across a parallel-plate capacitor, and then removed, leaving charges on the plates as shown:

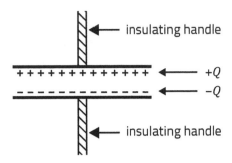

insulating handle

$+Q$

$-Q$

insulating handle

The plates are then pulled further apart, doubling their separation.

(a) Determine by what factor the energy stored in the capacitor is increased, stating clearly the principle on which your answer is based. [3]

....

....

....

....

....

(b) State where the extra energy has come from. [1]

....

Q5 Show that, for a given electric field strength, E, between the plates of a parallel-plate capacitor, the energy stored is proportional to the *volume* of the space between the plates. [2]

....

....

....

....

Q6 Write the *charges* (sign and magnitude) on the capacitor plates in the boxes provided. The space below the diagram is for your working. [4]

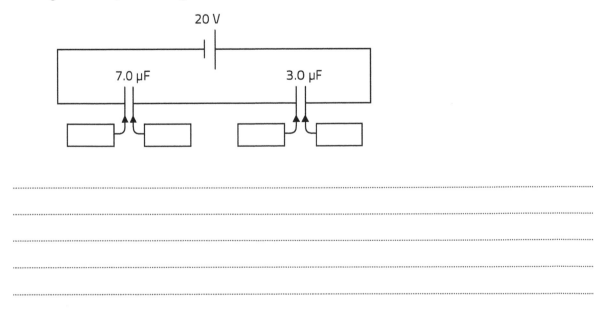

20 V

7.0 μF

3.0 μF

....

....

....

....

....

Q7 A capacitor combination is shown:

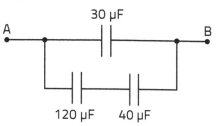

(a) Calculate its capacitance. [3]

....

....

....

....

(b) A battery of emf 12 V is connected across AB. Calculate:

(i) The charge that flows through the battery while the capacitors charge. [1]

....

(ii) The final pd across the 120 µF capacitor, giving your reasoning. [2]

....

....

....

Q8 A battery is connected across a 50 mF capacitor, C_1. The battery is then disconnected, leaving a pd of 9.0 V between the plates of C_1. An (initially) uncharged 50 mF capacitor, C_2, is now connected across C_1.

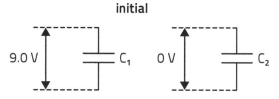

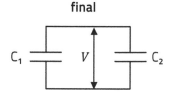

(a) Explain why the 'final' pd, V, is 4.5 V. [3]

....

....

....

....

....

(b) Calculate the change in total energy stored when C_2 is connected across C_1. [3]

....

....

....

....

....

Q9 In the circuit shown, the capacitor is initially uncharged. The switch is closed at time $t = 0$.

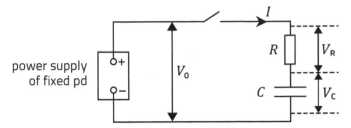

After time $t = 0$ the following equation applies:

$$Q = Q_0(1 - e^{-t/RC})$$

(a) What is the meaning of Q_0? [1]

....

(b) In each of the following parts, use the preceding equation to show that:

(i) $$V_C = V_0(1 - e^{-t/RC})$$ [2]

....

....

....

(ii) $$V_R = V_0 e^{-t/RC}$$ [2]

....

....

....

(iii) $$I = I_0 e^{-t/RC}$$ [2]

....

....

....

(c) (i) Starting from definitions of capacitance and resistance, show that the SI unit of Q_0/RC is the ampère. [3]

....

....

....

(ii) Explain why the gradient of the tangent at $t = 0$ to a graph of Q against t is Q_0/RC. [2]

....

....

....

Q10 The pd, V_c, between the plates of a capacitor is plotted against time as the capacitor discharges through a resistor.

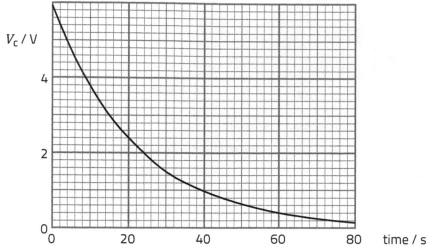

(a) Draw a circuit diagram of the arrangement that might have been used to obtain the results. Include a means of charging the capacitor beforehand. [3]

(b) (i) Using three points on the graph, verify that V_c decays exponentially with time. [3]

....

....

....

....

....

(ii) Determine the *time constant* of the decay. [2]

....

....

....

(c) For a graph of ln $(V_c$/volt) against time, give the values of the gradient and the intercept on the ln $(V_c$/volt) axis. [2]

....

....

....

Question and mock answer analysis

Q&A 1 Lauren investigated the energy stored in a capacitor using the apparatus shown:

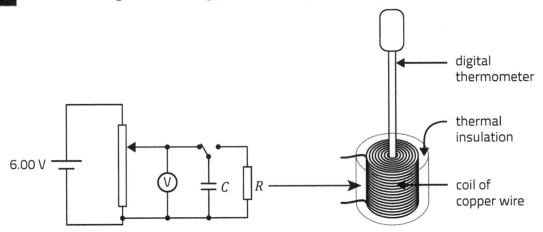

Lauren read the temperature, θ_1, of the coil of copper wire. She then charged the capacitor to a pd, V, and connected it across the coil (R in the circuit diagram). She read the coil's temperature, θ_2, after the capacitor had been discharging for 30 s. She repeated the procedure for six more pds, up to 5.0 V, and then again for all seven pds. Her plot of $(\theta_2 - \theta_1)$ against V^2 is given.

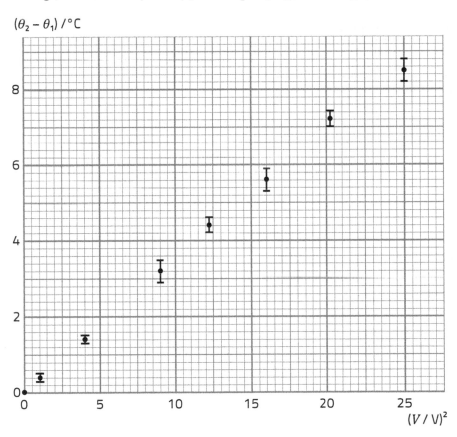

(a) Name the part of the circuit that enabled Lauren to select the various pds. [1]

(b) Suggest how Lauren decided on the vertical position of her points, and the error bars that she used. [2]

(c) Lauren's decision to plot $(\theta_2 - \theta_1)$ against V^2 was based on the equation:

$$\tfrac{1}{2} CV^2 = mc\,(\theta_2 - \theta_1)$$

in which C = capacitance of capacitor = 5.00 F

m = mass of copper in coil = 15.8 g

c = specific heat capacity of copper = 385 J kg^{-1}°C^{-1}

(i) State what quantities are represented by $\tfrac{1}{2} CV^2$ and $mc\,(\theta_2 - \theta_1)$. [1]

(ii) Evaluate to what extent Lauren's data, as plotted, support the equation. [7]

(d) (i) Decide, using a suitable calculation, whether or not 30 s was a long enough time to allow for capacitor discharge. [Resistance of coil = 2.0 Ω] [3]

(ii) Suggest why allowing a much longer time (for example 5 minutes) before reading the temperature, θ_2, would be likely to cause error. [1]

What is being asked

This question is centred on a specified practical investigation of the energy stored in a capacitor. It is not assumed that you will be familiar with exactly the same apparatus as that described, but enough information is given for you to figure out how it works. The question (like the investigation itself) is synoptic, as electric circuits and thermal physics are both touched upon.

Part (a) is a straightforward interpretation of the diagram, AO1; part (b) tests your familiarity with using experimental data, applied to these results, hence AO2. Part (c)(i), AO1, tests whether you understand the physical principle on which the equation is based, essentially recognition of two energy terms. In (c)(ii) you are left to your own devices to decide on your strategy, which makes it AO3, calling for calculations and conclusions. In (d), part (i) is another evaluation, AO3, with more than one way of answering it for three marks, but the single mark in part (ii) suggests there is really only one point to be made.

Mark scheme

Question part		Description	AOs 1	2	3	Total	Skills M	P
(a)		Potential divider. Accept 'potentiometer' [1]	1			1		1
(b)		Point plotted at mean of the two temp readings [1]		2		2		2
		Error bar runs between the two temperature readings [1]						
(c)	(i)	Energy [initially] in capacitor; gain in internal energy [accept thermal energy, heat, random energy] of coil [finally]	1			1		
	(ii)	At least one straight line through origin drawn, passing through all error bars, or all except the one furthest from the origin [1]			7	7	3	6
		Comment made that data points fit straight line through origin, as equation predicts [1]						
		but the point furthest from origin is anomalous [1]						
		Gradient of any straight line (e.g. attempt at best fit, or maximum gradient) correctly calculated for the chosen line [1]						
		Maximum gradient stated to be = 0.37 or 0.35 [°C V^{-2}] [1]						
		Theoretical gradient = 0.41 [°C V^{-2}] [1]						
		Comment that even maximum gradient is too small, so [to this extent] data don't fit equation [1]						
(d)	(i)	Time constant [= 5.0 F × 2.0 Ω] = 10 s [1]			3	3	1	3
		Either						
		[e^{-3}=] 5.0% of original charge (or voltage/pd) or 2.5% of energy left after 30 s [1]						
		so long enough **or** but not long enough [1]						
		Or						
		30 s is [considerably] greater than time constant therefore long enough **or** but not long enough [1]						
		[2 marks maximum for this approach]						
	(ii)	Heat will escape through the thermal insulation [in such a long time] [1]		1		1		1
Total			2	3	10	15	4	13

Rhodri's answers

(a)   A variable resistor ✗

MARKER NOTE
A variable resistor is a 2-terminal device. The 3-terminal device shown is a (variable) potential divider.   **0 marks**

(b)   Point is plotted midway between the two temperatures measured for that voltage. ✓
Error bar represents the uncertainty. ✗ [Not enough]

MARKER NOTE
Rhodri is right about the position of the point, but his comment about error bars is too vague.   **1 mark**

(c) (i) The left-hand side is the energy stored in the capacitor. The right-hand side is the heat given to the copper wire coil. ✓

MARKER NOTE
Rhodri is misusing the term 'heat', which should be reserved for the transfer of energy due to a difference in temperature, but the mark scheme generously allows it.   **1 mark**

(ii)

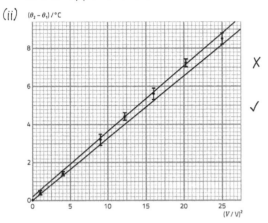

Points almost fit a straight line through the origin. This is right because equation says that $(\theta_2 - \theta_1)$ proportional to V^2. ✓

Least gradient $= \dfrac{8.2}{25} = 0.33°\text{C V}^2$ ✓

Largest $= (8.8 - 0.2)/25 = 0.34°\text{C V}^2$

In theory, gradient $= \dfrac{(C/2)}{mc}$

$= 2.5 \div 6.08$

$= 0.41\ °\text{C V}^2$. ✓

MARKER NOTE
Neither of Rhodri's lines goes through the origin and through all but one of the error bars, so he misses the 1st mark. But his comment is correct giving him the 2nd mark, though he has not mentioned the anomalous point, so misses the 3rd mark. He has correctly calculated the gradient of a chosen line (✓) but his maximum gradient is out of tolerance because his line started 'above' the origin. He's worked out the 'theoretical' gradient correctly (✓) but has not commented on the discrepancy.   **3 marks**

(d) (i) $CR = 5 × 2 = 10\text{ s}$ ✓

30 s is 3 times as long, so the capacitor will be well discharged. ✓

MARKER NOTE
Rhodri has realised the key significance of the time constant, has calculated it correctly and made a sensible comment. For 3 marks, though, he needed to give an idea of the fraction of the total energy not accounted for.   **2 marks**

(ii) Perhaps the coil will have started to cool off! ✓

MARKER NOTE
Not what the mark scheme says, but Rhodri's answer is equivalent.   **1 mark**

Total   **8 marks /15**

Ffion's answers

(a)   Variable potential divider ✓

(b)   Ends of error bar are at the two temperatures. ✓
Point is plotted at their mean. ✓

(c) (i) $CV^2/2$ is the energy in the charged capacitor.
$mc(\theta_2 - \theta_1)$ is the internal energy in the
copper after the discharge. ✗

(ii)

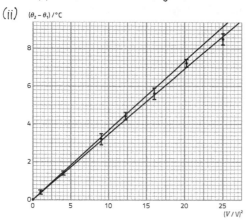

Rearranging equation:
$(\theta_2 - \theta_1) = \dfrac{C}{2mc} V^2$ so we should have a straight
line through the origin, ✓ of gradient

$\dfrac{C}{2mc} = \dfrac{5.0}{(2 = 0.0158 = 385)} = 0.41$ ✓

The points, all except the last one ✓, do seem to
fit a straight line through the origin ✓. Ignoring
the last point,

Lowest gradient $= \dfrac{8.7}{25} = 0.348$ ✓

Highest gradient $= \dfrac{9.2}{25} = 0.375$ ✓

So even the steepest line has too small a gradient,
so the fit is not so good after all. ✓

(d) (i) $Q = Q_0 e^{-t/RC}$ so $Q = Q_0 e^{-30/10} =$ ✓ 0.050

Therefore $\dfrac{V}{V_0} = \dfrac{1}{20}$ and $\left(\dfrac{V}{V_0}\right)^2 = \dfrac{1}{400}$, ✓
so almost no energy left in the capacitor. ✓

(ii)   The coil will have had time to lose heat
to the surroundings, so the temperature
rise will be lower than predicted. ✓

Section 5: Solids under stress

Topic summary

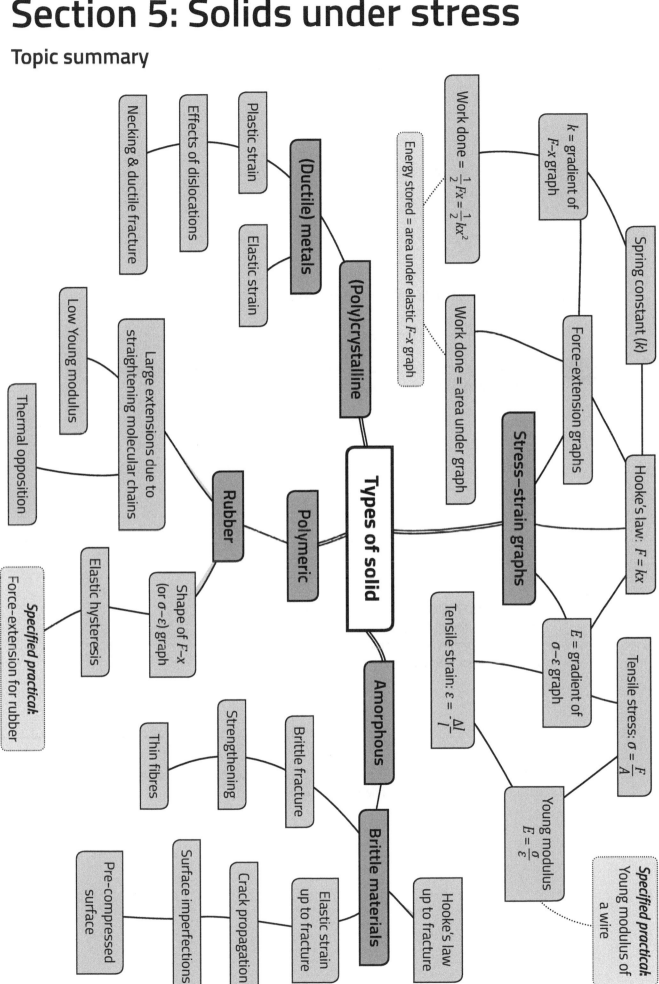

Q1 Hooke's law for a spring can be expressed by the equation:

$$F = kx$$

(a) State the meaning of the symbols, F, k and x, in this equation. [1]

F ....

k ....

x ....

(b) Express the unit of k in terms of the base SI units. [2]

....

....

....

(c) State the condition for this equation to be valid. [1]

....

Q2 Solid materials can be classified as *crystalline, amorphous* or *polymeric*. State what each of these terms means and give an example of each type of material. [3]

....

....

....

....

....

....

Q3

(a) Use the axes to sketch stress–strain (σ–ε) graphs for a brittle material, such as glass, and a ductile material, such as copper. Label the graphs. [2]

(b) Brittle materials are weak under tension. Describe one way of increasing the tensile strength of a brittle material and explain briefly how it works. [3]

....

....

....

....

....

Q4 In an experiment to determine a spring constant, Aled measured the extension of a spring when a load of mass (300 ± 6) g was hung from it. His result was (15.1 ± 0.2) cm.

(a) Calculate the value of the spring constant together with its **absolute** uncertainty. [4]

....

....

....

....

....

....

(b) A year-13 student told Aled that the period, T, of vertical oscillation of a mass, m, on a spring is related to the spring constant, k, by the equation:

$$T = 2\pi\sqrt{\frac{m}{k}}$$

Design a way in which Aled could use the result of his experiment in part (a) and another measurement to test this equation. [3]

....

....

....

....

....

....

Q5 Some materials are *ductile*.

(a) State what is meant by a ductile material. [1]

....

....

(b) Ductile metals are *polycrystalline*.

(i) State what polycrystalline means. [1]

....

....

(ii) Explain how the presence of dislocations accounts for the ductile nature of polycrystalline metals. A diagram may help your explanation. [3]

....

....

....

....

....

....

Q6 Ductile metals show *elastic strain* at low values of *stress*. Above the *elastic limit*, they exhibit *plastic strain*. State the meaning of the italicised terms. [4]

....

....

....

....

....

....

Q7 Wire **A** has length l and diameter d and is made from a metal with Young modulus E. It is joined at the end to wire **B** which has length $2l$, diameter $2d$ and Young modulus $1.5E$. A force, F, is applied to the end of each wire as shown.

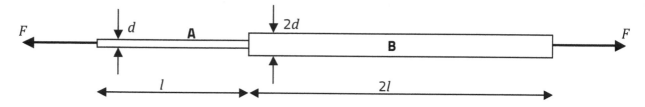

In the question that follows, σ_A is the strain in wire **A**, W_A is the work done in in extending wire **A**, etc. Complete the table to show the ratios of the given quantities. [6]

Quantity	Ratio
Tension	$\dfrac{F_A}{F_B} =$
Stress	$\dfrac{\sigma_A}{\sigma_B} =$

Quantity	Ratio
Strain	$\dfrac{\varepsilon_A}{\varepsilon_B} =$
Extension	$\dfrac{\Delta l_A}{\Delta l_B} =$

Quantity	Ratio
Work	$\dfrac{W_A}{W_B} =$
Potential energy per unit volume	$\dfrac{W_A/V_A}{W_B/V_B} =$

Space for calculations:

Q8 Joel used a piece of steel wire, of length 3.550 m, to obtain a value for the Young modulus, E, of steel. He put the wire under tension by hanging a load from its end and measured the diameter at various places. His results were:

Diameter /mm: 0.23 0.23 0.25 0.23 0.24 0.25

He added an extra load of 0.800 kg to the wire and measured the extra extension to be 3.1 mm.

(a) Why did Joel hang a load from the end of the wire before measuring the diameter? [1]

....

(b) Estimate the **percentage** uncertainty in Joel's value for E, explaining your reasoning. There is no need to calculate E from Joel's results. [3]

....

....

....

....

....

(c) Bethan said that using a thinner piece of wire would reduce the uncertainty in the result. Discuss whether she was correct. [2]

....

....

....

....

Q9 A steel cable has a diameter of 3.0 cm and a length of 5.0 km. The steel has a *yield stress* of 300 MPa and a Young modulus of 2.0 GPa.

(a) State the meaning of the term in italics. [1]

....

....

(b) In use, the maximum safe working stress is one fifth of the yield stress. Calculate the elastic potential energy stored in the cable with this stress. [3]

....

....

....

....

....

....

Component 2 Practice questions

Q10 The string of a modern longbow, used in archery contests, applies a maximum force of 280 N to a 50 g arrow, for a draw length (i.e. the distance the arrow is pulled backwards) of 76 cm.

(a) Assuming that the force is proportional to the distance the arrow is drawn back, calculate the work done in drawing the bow. [2]

....

....

....

(b) The efficiency of energy transfer to the arrow has been estimated as 90%. A sales brochure states that the bow is capable of shooting an arrow a distance of 400 m, if the arrow is shot at an angle of 45° to the horizontal. Evaluate this claim. [You should neglect the effects of air resistance.] [5]

....

....

....

....

....

....

....

....

Q11 It is proposed to place a set of energy-absorbing buffers at the end of a local railway line, to stop slow-moving trains from overshooting the railway platform.

The buffers contain springs which are compressed and absorb the energy of the trains. Two designs are proposed with different spring constants. Evaluate the advantages and disadvantages of using a spring with a lower spring constant. [4]

....

....

....

....

....

....

....

Question and mock answer analysis

Q&A 1 A student investigates the tensile properties of a rubber band of original length 10 cm. She produces this graph:

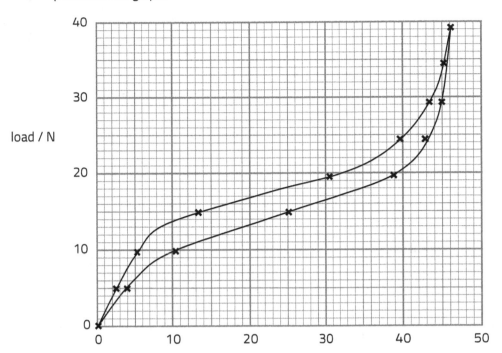

(a) Use the graph to help explain the terms *elastic* and *hysteresis*. [3]

(b) The crosses represent the results obtained by the student. Suggest briefly how the student carried out the investigation, including how the values of force were obtained. [3]

(c) Another student says that the results are not satisfactory between loads of 5 N and 15 N. Explain this comment and suggest an improvement to the method. [2]

(d) Rubber is different from metals in that large strains are possible and low tensile forces are involved. Account for these differences in terms of molecular structure. [4]

(e) Estimate the energy dissipated in loading and unloading the rubber band, explaining how you obtained your answer. [3]

What is being asked

This is a question centred on a specified practical. It has aspects of experimental design – parts (b) and (c) – the recall of an explanation of the mechanical properties of rubber – part (d) – and interaction with the data – parts (a), (b) and (e).

Mark scheme

Question part	Description	AOs 1	AOs 2	AOs 3	Total	Skills M	Skills P
(a)	An elastic material returns to its original size and shape when a stress is removed. [1] <u>In this case there is zero extension at the end of the unloading curve.</u>	1			3		
	In elastic hysteresis, load-extension graphs for unloading and loading do not coincide. [1] <u>In this case the unloading curve is underneath the loading curve.</u>	1					
	Both underlined sections. [1]		1				

Question part			Description	AOs			Total	Skills	
				1	2	3		M	P
(b)			The student had a set of loads (or masses) which she added to the rubber band up to the maximum, measuring the extension each time. [1]	2	1		3	1	3
			She repeated the measurements whilst decreasing the load (in the same way). [1]						
			Numerically relating weight to masses, e.g. 3.8 N is produced by a mass of 4.0 kg. [1]						
(c)			On the loading curve, there is a marked change of slope between 5 N and 15 N (so the shape of the curve is uncertain). [1]		2		2		2
			Obtain results for different masses (e.g. 0.7, 0.9, 1.2 kg) between. [1]						
(d)			Extensions in other materials, e.g. steel, are produced by stretching very stiff [inter-atomic] bonds. [1] These bonds are short range and so the material breaks at small strains [even if the material is ductile]. [1]	4			4		
			In rubber, large extensions are produced by the straightening out of long C–C chains [1] [which is possible by the rotation of the single bonds], and needs much lower forces. [1]						
(e)			Identification of the area between the curves as the dissipation. [1]	1			3		
			Reasonable method attempted, e.g. counting squares or division into geometrical shapes. [1]		1				
			Answer in range 1.3–1.7 J [1]		1			1	
Total				9	4	2	15	2	5

Rhodri's answers

(a) Elastic materials go back to their original shape. **X** (no bod)

Hysteresis is where there is a loop between the two graphs. **✓** (bod)

MARKER NOTE

Rhodri doesn't achieve the first marking point; he should have said that the original shape and size were regained when the stress was removed, or words to that effect. In the second, the expression 'loop between the graphs' is slightly unclear but effectively is the same as required and the examiner awards the mark, by bod.

No reference is made to the actual graph so the third marking point is not awarded.

1 mark

(b) The loads are multiples of just less than 5 N, which is the weight of a 0.5 kg weight. **✓** The student added 0.5 kg weights one by one and measured the extension, **✓** then removed the loads one after the other and measured the extension. **✓**

MARKER NOTE

Rhodri has correctly observed that the increments of load were the weight of a 0.5 kg mass. It might have been nicer if he had used $W = mg$ to show this more explicitly but he has done enough to obtain the mark. He has proceeded to relate this to the standard method of this experiment for the other two marks.

3 marks

(c) He should repeat the readings and take an average to make sure they are more accurate (or check the accuracy). **X**

MARKER NOTE

Rhodri has failed to spot that this question was about the actual results obtained. He has given the weak answer of 'repeat and average' when this will not solve the problem in this case. More data are needed to fill in the gap and hence determine the precise shape of the graph here.

0 marks

(d) This rubber band had an extension of 46 cm for an original length of 10 cm and the load was only 38 N. If it had been steel, the extension would have been a lot less – just a few mm. This is because rubber is made of long-chain molecules, which can stretch out. ✓

> **MARKER NOTE**
>
> For most of his answer, Rhodri has just repeated the question, albeit with the addition of some numerical data. The only new aspect, for which he obtains credit, is the unravelling of tangled molecules. He has not explained why steel is stiff or why it can only stretch a short distance. He should also have explained why large extensions for rubber were achieved with low forces.
>
> **1 mark**

(e) Average distance between graphs = 3 N (by eye)

Distance stretch = 46 cm = 0.46 N ✓

∴ Area between graphs = 0.46 × 3 = 1.38 J

Say 1.4 J ✓

> **MARKER NOTE**
>
> A nice way of estimating the 'area' between the curves (for a mark) and an answer well within the expected range (for another). Rhodri misses the easiest mark by not mentioning the significance of this area in terms of the dissipation of energy.
>
> **2 marks**

Ffion's answers

(a) The graph shows that, when the load is removed, the extension goes back to zero. This is what is meant by elastic. ✓

The graph shows that the load-extension graphs for loading and unloading are different ✓ – this is elastic hysteresis. ✓

> **MARKER NOTE**
>
> This was a good answer. Ffion correctly explains both *elastic* and *hysteresis* and identifies the features of the graph which illustrate these properties.
>
> A good concise answer.
>
> **3 marks**

(b) The student had a set of weights up to 3.8 N. She put them on, one after the other, and measured the extension from the beginning each time. ✓ She then removed the weights one by one and measured the extensions (from the beginning) again. ✓

> **MARKER NOTE**
>
> Ffion has correctly described the procedure for loading and unloading, together with determining the extension for this specified practical, and hence obtained marks two and three. She identified the 3.8 N maximum load but doesn't relate this to the masses needed to achieve this weight and so misses the numerical part of the question.
>
> **2 marks**

(c) There should really be more results. It's a complicated graph, so to be sure you need more readings. ✓ bod, ✗

> **MARKER NOTE**
>
> Ffion has something of the right idea. The examiner has awarded a mark bod. for *more data* because of the phrase 'complicated graph'. To obtain the second mark, Ffion should have gone on to explain the nature of the complications, i.e. large change of gradient, and the need for more data in the specified area.
>
> **1 mark**

(d) Rubber is a tangle of long molecules. These can be straightened out a lot, so large extensions are possible. ✓ This doesn't take much force because bonds are not being stretched – just rotated. ✓

Metals are different because the bonds between the atoms need stretching which takes much bigger forces. ✓

> **MARKER NOTE**
>
> Ffion obtains both of the marks available for the properties of rubber – extensible and low tension needed – and obtains the marks.
>
> Her answer as to why large forces are needed to stretch steel is correct and elicits a mark. She makes no attempt at an explanation of the fact that steel can only extend a few percent before fracture occurs, which is needed for the fourth mark.
>
> **3 marks**

(e) Area under graph is the work done:

To find this, count the squares:

A 1 cm square = 0.05 m × 5 N = 0.25 J ✓

Stretching: Total number of squares = 31

∴ Work done = 7.75 J ✓

Contracting: 25 squares → 6.25 J

∴ Energy lost as heat = 7.75 − 6.25 = 1.5 J ✓

> **MARKER NOTE**
>
> Ffion quite reasonably estimates the work done (on the rubber band) in stretching the rubber band, by square counting. She also estimates the work done (by the band) in contracting. She recognises, and states, that the difference between these is the 'loss' in energy as 'heat', which is acceptable at AS and picks up all three marks.
>
> **3 marks**

Total	12 marks / 15

Q&A 2 Briefly describe the process by which a brittle material breaks under tension. [3]

What is being asked

This is a straightforward AO1 question which asks candidates to reproduce knowledge which is required by the specification. It is likely to be asked as part of a longer question.

Mark scheme

Description	AOs			Total	Skills	
	1	2	3		M	P
Cracks or imperfections in [surface of] material [1] Stress is concentrated at the tips of cracks, exceeding breaking stress [locally] [1] Cracks extend [further concentrating stress] [1]	3			3		
	3			**3**		

Rhodri's answers

Brittle materials break because of cracks or scratches on their surface. ✓ An example is builders `cutting` bricks by making a crack across them and hitting them. ✗

MARKER NOTE

Rhodri has correctly stated that brittle fracture is initiated by cracks on their surface and gains the first mark. Instead of giving more details of the process, which would gain him more marks, he proceeds to give an example, which is not required.

1 mark / 3

Ffion's answers

Brittle materials, such as glass, have small holes or cracks in them. ✓ When the material is stretched, the stress at the edge of the cracks is much bigger than the average stress. If this stress is big enough, the bonds at the end of the crack break, ✓ which makes the crack grow until it extends across the material. ✓ So the material breaks.

MARKER NOTE

Ffion has given a good account of the process of brittle fracture. She has identified the significance of cracks for the first mark and correctly stated that the stress at the tip of a crack is magnified. She talks correctly of this stress breaking bonds, for which the examiner awards the second mark. Ideally she would have said that the stress at the new tip of the crack is now even greater, leading to runaway crack growth, but the marking scheme does not require this and she is awarded the third mark for the statement that the crack grows..

3 marks / 3

Section 6: Electrostatic and gravitational fields of force

Topic summary

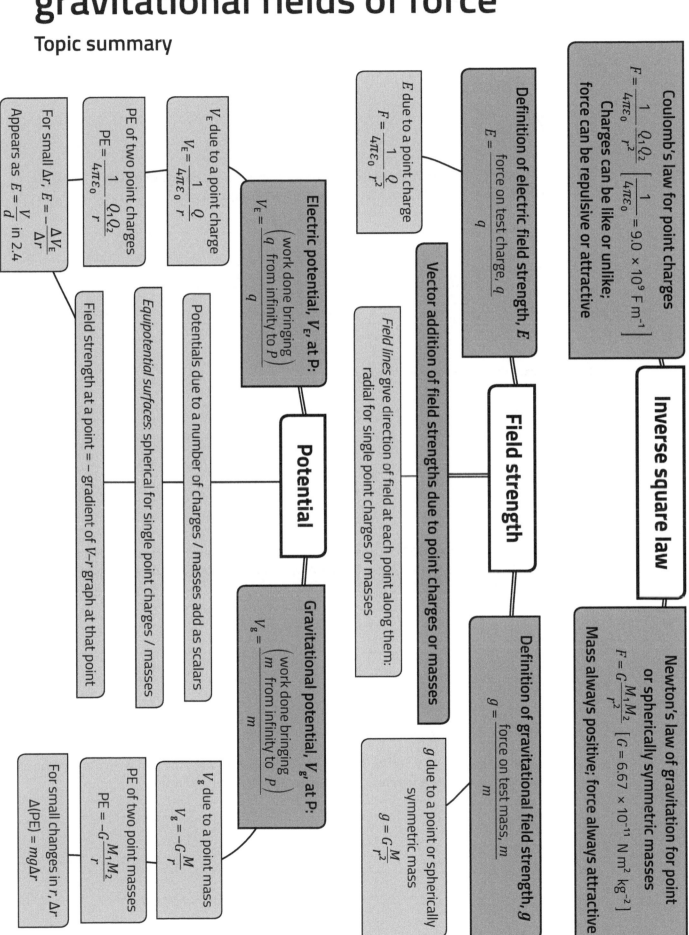

Coulomb's law for point charges

$$F = \frac{1}{4\pi\varepsilon_0}\frac{Q_1 Q_2}{r^2} \quad \left[\frac{1}{4\pi\varepsilon_0} = 9.0 \times 10^9 \text{ F m}^{-1}\right]$$

Charges can be like or unlike; force can be repulsive or attractive

Inverse square law

Newton's law of gravitation for point or spherically symmetric masses

$$F = G\frac{M_1 M_2}{r^2} \quad [G = 6.67 \times 10^{-11} \text{ N m}^2 \text{ kg}^{-2}]$$

Mass always positive; force always attractive

Definition of electric field strength, E

$$E = \frac{\text{force on test charge, } q}{q}$$

E due to a point charge

$$F = \frac{1}{4\pi\varepsilon_0}\frac{Q}{r^2}$$

Vector addition of field strengths due to point charges or masses

Field strength

Definition of gravitational field strength, g

$$g = \frac{\text{force on test mass, } m}{m}$$

g due to a point or spherically symmetric mass

$$g = G\frac{M}{r^2}$$

Field lines give direction of field at each point along them: radial for single point charges or masses

Electric potential, V_E, at P:

$$V_E = \left(\frac{\text{work done bringing } q \text{ from infinity to } P}{q}\right)$$

V_E due to a point charge

$$V_E = \frac{1}{4\pi\varepsilon_0}\frac{Q}{r}$$

PE of two point charges

$$PE = \frac{1}{4\pi\varepsilon_0}\frac{Q_1 Q_2}{r}$$

For small Δr, $E = -\frac{\Delta V_E}{\Delta r}$
Appears as $E = \frac{V}{d}$ in 2.4

Field strength at a point = $-$ gradient of V–r graph at that point

Equipotential surfaces: spherical for single point charges / masses

Potentials due to a number of charges / masses add as scalars

Potential

Gravitational potential, V_g, at P:

$$V_g = \left(\frac{\text{work done bringing } m \text{ from infinity to } P}{m}\right)$$

V_g due to a point mass

$$V_g = -G\frac{M}{r}$$

PE of two point masses

$$PE = -G\frac{M_1 M_2}{r}$$

For small changes in r, Δr
$$\Delta(PE) = mg\Delta r$$

Q1 Two identical small spheres, each of mass 2.00×10^{-7} kg, carry equal positive charges and hang on insulating threads from a fixed point. When in equilibrium they hang as shown.

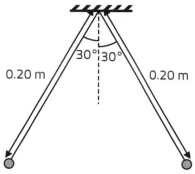

(a) Show that the electrostatic repulsive force on each sphere must be approximately 1.1×10^{-6} N. Use the space to the right above for a vector diagram if required. [3]

....

....

....

....

....

(b) Calculate the charge on each sphere. [3]

....

....

....

....

....

Q2 (a) The Moon's radius is 1737 km, and the gravitational field strength at its surface is 1.62 N kg^{-1} towards its centre. Show that the Moon's mass is approximately 7×10^{22} kg, stating an assumption that you make. [3]

....

....

....

....

....

(b) The Earth's mass is 5.97×10^{24} kg, and the mean distance between the centres of the Earth and Moon is 3.84×10^{8} m. Calculate the gravitational pull of the Moon on the Earth. [2]

....

....

....

....

Q3 (a) The mass of a proton is 1.67×10^{-27} kg. For two protons a given distance apart, calculate the ratio:

$$\frac{\text{electrostatic force between protons}}{\text{gravitational force between protons}}$$

[3]

....

....

....

....

....

(b) Protons make up a sizeable fraction (by mass) of both the Sun and the Earth. Explain why the gravitational force between these bodies is far greater than the electrostatic force. [3]

....

....

....

....

....

Component 2 Practice questions

Q4 (a) In the space on the left below draw eight field lines showing the electric field around an isolated negative charge. [2]

Diagram for part (a)
(to be completed)

Diagram for part (b)

(b) Some field lines in the area around equal and opposite point charges are shown on the right. By considering the fields due to individual charges, explain why the field line at P is in the direction shown. [3]

....

....

....

....

....

Q5 Equal and opposite charges are separated by a distance of 0.12 m, as shown:

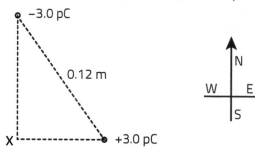

(a) (i) Show clearly that the electric field strength at **X** due to the negative charge is 2.8 N C^{-1} Northwards. [3]

....

....

....

....

(ii) Determine the electric field strength at **X** due to the positive charge. [2]

....

....

....

(iii) Determine the resultant field strength at **X**. Use the space to the right of the diagram for a vector diagram, if required. [3]

....

....

....

....

(b) (i) Calculate the potential at point **X**. [3]

....

....

....

....

(ii) A proton is released from point **X**. Calculate the maximum *speed* that it attains, giving your reasoning. [Proton mass = 1.67×10^{-27} kg] [3]

....

....

....

....

Q6   Three equipotential surfaces are shown for an isolated positive point charge, Q.

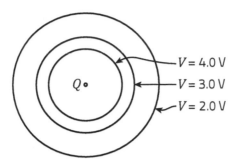

Explain why:

(a) The surfaces are spherical.   [1]

....

....

(b) The surfaces are not equally spaced apart, even though their potentials differ by the same size step
     (1.0 V).   [3]

....

....

....

....

....

Q7 (a) Explain why gravitational potentials are given as negative quantities. [The explanation is not the
             negative sign in $V = -GM/r$.]   [2]

....

....

....

(b) A rocket is launched vertically from Mars at a velocity of 3000 m s^{-1}.

(i)   Explain why it is necessary to use the equation, $PE = -GMm/r$, rather than the equation
      $\Delta(PE) = mgh$, in relation to the rocket's subsequent motion.   [2]

....

....

....

(ii) Calculate the maximum height above the Martian surface the rocket will reach. [4]
      [Mass of Mars = 6.42×10^{23} kg; Diameter of Mars = 6780 km]

....

....

....

....

....

Q8 Point charges, $+Q$ and $-Q$, are placed a distance $2d$ apart, as shown:

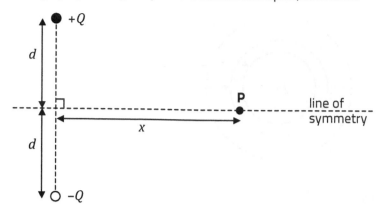

(a) (i) Determine the *electric field strength* at point P, in terms of Q, ε_0, d and x. You may add to the diagram. [4]

....

....

....

....

....

....

....

(ii) Adam says that the potential at P is zero. Bethan says that this can't be right because work has to be done bringing a test charge along the line of symmetry from infinity to P. Evaluate who is right, explaining why the other is wrong. [3]

....

....

....

....

....

(b) **Compare** the variation in the electric field strength for the +/- charges with distance (from zero up to a large distance) **with** the case of two equal positive charges arranged in the same way. [4]

....

....

....

....

....

....

....

Question and mock answer analysis

Q&A 1

(a) State what is meant by the *gravitational potential* at a point. [2]

(b) Two stars, A and B (see diagram), are much closer to each other than to any other stars.

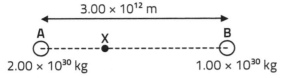

In the graph the gravitational potential, V, along a line joining A and B is plotted against r_A, the distance from star A. The distance **AX** is 1.00×10^{12} m.

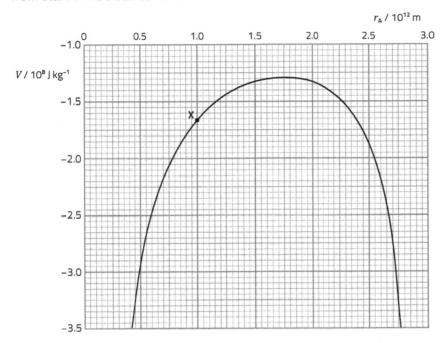

(i) Using data from the diagram, show clearly that point **X** on the graph is correct. [3]

(ii) **Use the graph** to determine the **resultant field strength** at **X**. [4]

(iii) Determine **from the graph** the value of r_A for which the resultant gravitational field strength is zero, giving a brief justification. [2]

(iv) Explain **in terms of field strength or force** why the point at which the resultant field strength is zero is closer to B than to A. Calculations calling for the use of a calculator are not needed. [2]

What is being asked

Part (a) serves two purposes: it tests that you know a standard AO1 definition and it sets the scene for part (b).

Part (b) consists mainly of AO2 as you are applying given data.

In (b)(i), rather than asking you to plot your own graph (taking several minutes of your time) the examiner finds out just as much about your understanding of the physics involved by asking you to check a single point. You must, though, present your check clearly.

(b)(ii) requires the use of a specific relationship and a specific technique. You must use the graph (as instructed in bold!) even if another method occurs to you.

In (b)(iii) it's a pretty good guess that one of the marks will be for the 'brief justification' leaving only one mark for the actual value, so it can't require much labour to determine!

(b)(iv) departs from *potential*, the main theme of the question. Although numerical calculations are not needed, you are allowed to use equations. Words will also be required!

Mark scheme

Question part		Description	AOs			Total	Skills	
			1	2	3		M	P
(a)		The work done [by an external force] taking a [test] mass from infinity to the point [1] divided by the [test] mass. [1]	2			2		
(b)	(i)	$V_A = -\dfrac{6.67 \times 10^{-11} \times 2.00 \times 10^{30}}{1.00 \times 10^{12}}$ [1] $= -1.334 \times 10^8$ J kg^{-1} $V_B = -\dfrac{6.67 \times 10^{-11} \times 1.00 \times 10^{30}}{2.00 \times 10^{12}}$ [1] $= -0.334 \times 10^8$ J kg^{-1} $V_A + V_B = -1.67 \times 10^8$ J kg^{-1} ecf **and** comment that this agrees with graph [or, to obtain ecf credit, doesn't agree]. [1]		3		3	1	
	(ii)	Tangent drawn to curve at **X** [1] In gradient equation 'rise' and 'run' put in correctly, but tolerating slips in powers of 10. [1] g from 1.1×10^{-4} to 1.3×10^{-4} N kg^{-1} **unit** [1] g stated to be towards **A**. Accept to the left. [1]		4		4	4	
	(iii)	$r_A = 1.75 \times 10^{12}$ [m] [$\pm 0.05 \times 10^{12}$ m] [1] Gradient is zero here, or turns from + to −. [1]	1	1		2	1	
	(iv)	For zero resultant field: $\dfrac{[G]M_A}{r_A^2} = \dfrac{[G]M_B}{r_B^2}$ **or** inverse sq. law referred to [1] $M_B < M_A$ so $r_B < r_A$ **or** clear argument in words [1]		2		2		
Total			3	10	0	13	6	

Rhodri's answers

(a)   The work done when a body is taken from infinity to the point. ✓ X

MARKER NOTE
Rhodri has the main thrust of the definition correct, but the omission of 'per unit mass' is a serious mistake. He has, in effect, defined potential energy rather than potential.   **1 mark**

(b) (i)   V due to A $= -\dfrac{6.67 \times 10^{-11} \times 2.00 \times 10^{30}}{1.00 \times 10^{12}}$ ✓

$= -1.33 \times 10^8$

V due to B $= -\dfrac{6.67 \times 10^{-11} \times 1.00 \times 10^{30}}{2.00 \times 10^{12}}$ ✓

$= -0.33 \times 10^8$

$V_A - V_B = -1.00 \times 10^8$ X

The agreement with the graph (-1.7×10^8) is not very good X no bod

MARKER NOTE
Rhodri has calculated V_A and V_B correctly, but has *subtracted* V_B from V_A, probably confusing the scalar V with the vector g.   **2 marks**

(ii)

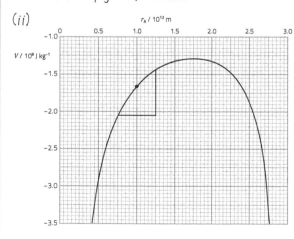

X X [No tangent drawn]

$$g = \frac{-1.45 \times 10^8 - (-2.05 \times 10^8)}{1.25 \times 10^{12} - 0.75 \times 10^{12}}$$

$$= 1.2 \times 10^{-4} \text{ N kg}^{-1} \checkmark \text{ bod}$$

Field is towards star A. $\checkmark$

> **MARKER NOTE**
> Rhodri has not drawn a tangent and loses the first two marks. The construction he has used produces an approximation to the gradient at 1.0×10^{12} m. He has actually calculated the <u>mean</u> value of g between 0.75 and 1.25×10^{12} m. In fact his value of g lies within the permitted range, and he has remembered to give the direction of this vector, so gains the last two marks.
> **2 marks**

(iii) The resultant field is zero when r_A is 1.75×10^{12} m $\checkmark$ because the graph is highest here. ✗

> **MARKER NOTE**
> Rhodri's value of r_A is correct, but his attempted justification doesn't connect with graph gradient. **1 mark**

(iv) Because star B is lighter, we need to be closer to it to get a field strength from it that is equal and opposite to the field strength from A. ✗ $\checkmark$

> **MARKER NOTE**
> Rhodri has a good feel for what's going on, and certainly deserves the 2nd mark, but he hasn't explained *why* being closer to B will compensate for B's smaller mass – hence he loses the 1st mark. **1 mark**

> **Total** **7 marks /13**

Ffion's answers

(a)   The work done per unit mass by the gravitational force when a mass goes from the point to infinity. $\checkmark\checkmark$

> **MARKER NOTE**
> Ffion's answer is equivalent to the standard definition given in the mark scheme. **2 marks**

(b) (i) $V_A + V_B =$
$$-6.67 \times 10^{-11} \times \left(\frac{2.00 \times 10^{30}}{1.00 \times 10^{12}} + \frac{1.00 \times 10^{30}}{2.00 \times 10^{12}} \right)$$
$$= 1.67 \times 10^8 \text{ J kg}^{-1} \checkmark\checkmark$$

This is exactly the value of V from the graph. $\checkmark$

> **MARKER NOTE**
> Ffion's working is clear and economical. She has clearly done this sort of calculation before! **3 marks**

(ii)

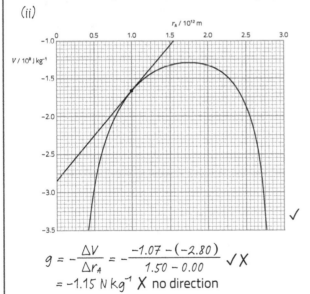

$$g = -\frac{\Delta V}{\Delta r_A} = -\frac{-1.07 - (-2.80)}{1.50 - 0.00} \checkmark ✗$$
$$= -1.15 \text{ N kg}^{-1} ✗ \text{ no direction}$$

> **MARKER NOTE**
> Ffion has drawn a good tangent. 1st mark gained. She has found the gradient correctly, apart from missing powers of 10. 2nd mark gained, 3rd lost. She has remembered that g is *minus* the potential gradient, but has not interpreted the minus sign in her answer in terms of a *direction*. 4th mark lost. **2 marks**

(iii) $r_A = 1.75 \times 10^{12}$ m $\pm 0.05 \times 10^{12}$ m $\checkmark$, as the gradient is zero somewhere in this region. $\checkmark$

> **MARKER NOTE**
> This is a very good answer. It was a nice touch to give an uncertainty, though there is no credit for it in the mark scheme. **2 marks**

(iv) For the field strengths from the two stars to cancel to zero, their magnitudes must be equal, but $g \propto m/r^2$, so for the same g, a smaller m requires a smaller r, that is the point of zero resultant is closer to B than A. $\checkmark\checkmark$

> **MARKER NOTE**
> Ffion's explanation is clearly and logically argued. **2 marks**

> **Total** **11 marks /13**

Section 7: Using radiation to investigate stars

Topic summary

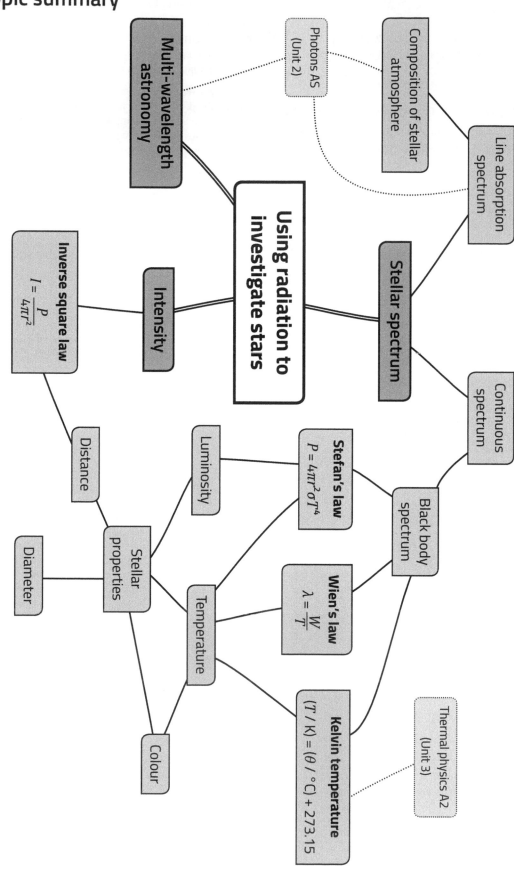

Q1 The joule (J) may be written in terms of the SI base units as kg m^2 s^{-2}. The SI unit for the *intensity* of radiation is W m^{-2}. Use defining equations to express this unit in SI base units. [3]

Q2 Define the term *black body* in terms of both the absorption and the emission of radiation. [2]

Q3 Red dwarf stars of the class M5V have radii approximately one quarter that of the Sun and a temperature (in kelvin) approximately half that of the Sun. The luminosity of the Sun is approximately 4×10^{26} W. Estimate the luminosity of M5V stars. [4]

Q4 The mass, M, and luminosity, L, of stars are often expressed in terms of the mass and luminosity of the Sun, $M_\odot$ and $L_\odot$ respectively. The greater the mass, the greater is its luminosity.

For masses in the range $0.43M_\odot < M < 2M_\odot$, the relationship in literature is often given as

$$\frac{L}{L_\odot} = \left(\frac{M}{M_\odot}\right)^4.$$

(a) The star 61 Cygni A has a mass of 0.70 $M_\odot$. Use the above equation to estimate its luminosity as a multiple of $L_\odot$. [1]

(b) In fact the luminosity of 61 Cygni A is 0.153 $L_\odot$. Alex says that the power of $(M/M_\odot)$ in the above equation should be less than 4. Evaluate whether he is right. [2]

Q5 The graph is of the continuous spectrum of the light from a distant star. The distance of the star from the Earth is known. The total intensity of the star's radiation arriving at the Earth is given by the area under the graph.

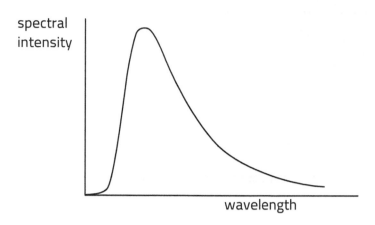

Explain how such a graph (with appropriate axis scales) can be used, together with the laws of radiation, to determine properties of the star. [6 QER]

....

....

....

....

....

....

....

....

....

....

....

....

....

Q6 The spectrum of a star consists of a *continuous emission spectrum* and a *line absorption spectrum*. Describe what is meant by these terms. [2]

....

....

....

....

Q7   Explain how information about a star's composition may be derived from its absorption spectrum.   [3]

....

....

....

....

....

Q8   The diameter of the Sun is 1.39×10^6 km and the wavelength of the peak emission is 501 nm.

(a)   Calculate the temperature of the solar surface.   [2]

....

....

....

....

(b)   A website gives the solar luminosity as 3.83×10^{26} W. Evaluate whether the data in this question are consistent with the Sun emitting radiation as a black body.   [3]

....

....

....

....

....

Q9   The surface of the Sun has a temperature of about 6000 K. Sunspots are small regions of the solar surface which have a temperature of about 4000 K.

(a)   (i)   Calculate the peak wavelength of the radiation emitted by a sunspot.   [2]

....

....

....

(ii)   Identify the region of the e-m spectrum in which the peak wavelength lies.   [1]

....

(b)   Sunspots appear black in images of the solar surface. Explain this. A calculation will help your answer.   [3]

....

....

....

....

....

Q10 Bryn says that, if Wien's law is correct, the photon energy of the peak of a star's spectrum is proportional to the temperature of the surface of the star. Evaluate whether he is correct. [3]

....

....

....

....

....

....

Q11 Describe what is meant by multiwavelength astronomy and give an example of its use. [2]

....

....

....

....

....

Q12 The table gives information about the temperature, T, of various sources of thermal radiation in the universe.

Source	T / K	Source	T / K
Cosmic microwave background radiation	2.7	Black hole inner accretion disk	10^7
Galactic molecular clouds (from which stars form)	10 – 50	Gas between galaxies (typical)	10^6
Red giant / Blue supergiant star surface	3 000 / 50 000	Supernova	5×10^7

Use data from this table to explain how multiwavelength astronomy is useful in studying the different processes which occur in the universe. [5]

....

....

....

....

....

....

....

....

....

....

Q13 The UV section of the e-m spectrum has a wavelength range of 10–400 nm. Calculate the range of photon energies in the UV spectrum. Express your answers in both J and eV. [4]

Q14 Singly ionised helium atoms, He$^+$ (i.e. helium atoms with one electron missing) are found in the atmospheres of hot stars. The bottom 6 energy levels in He$^+$ are:

–54.4 eV (ground state)   –13.6 eV   –6.0 eV   –3.4 eV   –2.2 eV   –1.5 eV

(a) An absorption line of wavelength 1.0 μm is observed in the infra-red spectrum of a star. Explain how this can arise from He$^+$ in the atmosphere of the star. [4]

(b) The energy of photons in the visible spectrum lies between 1.9 eV and 3.1 eV. Identify transitions in He$^+$ ions which can lead to dark lines in the visible absorption spectrum. [2]

(c) Eleri notices that there are no lines due to He$^+$ in the visible spectrum of the Sun. She claims that this is because the temperature of the Sun's surface is too low, about 6000 K. Evaluate this claim. [3]

Question and mock answer analysis

Q&A 1   To a good approximation, stars behave as *black bodies*.

(a) State what is meant by a black body in terms of both the emission and absorption of radiation. [2]

(b) State what is meant by the *luminosity* of a star. [1]

(c) Three of the characteristics of the Sun are as follows

Diameter = 1.4×10^6 km   Surface temperature = 5770 K   Luminosity = 3.83×10^{26} W

In the future, the Sun will pass through a stage in which its diameter is 300 million km and its surface temperature is 3000 K. It will then become a much smaller star with a diameter of 14 000 km and surface temperature 20 000 K.

Use the above information, together with calculations, to describe the changes in the appearance of the Sun to a distant observer. [5]

What is being asked

Like many questions, this one starts with some aspects of recall (AO1), designed to lead you into an application of physics (AO2). In this case, you should have learned the definitions of *black body* and *luminosity* to be able to answer parts (a) and (b). In answering part (c), you will need to identify the observational features which are amenable to calculation, using the Stefan-Boltzmann's law and Wien's law, do the relevant calculations and then interpret the results.

Mark scheme

Question part			Description	AOs			Total	Skills	
				1	2	3		M	P
(a)			[A black body is one which] absorbs all [electromagnetic] radiation which falls upon it ... [1]	2			2		
			...[and it] emits the most radiation possible at any wavelength [and that temperature]. [1]						
(b)			The luminosity of a star is the power emitted [or the energy emitted per unit time] in the form of electromagnetic radiation. [1]	1			1		
(c)			Use of Stefan's law attempted [even with slips], e.g. substitution in $L = A\sigma T^4$ (accept radius/diameter confusion or πr^2 used) at any stage or $\frac{L_1}{L_2} = \frac{A_1 T_1^4}{A_2 T_2^4}$ seen. [1]		5		5	4	
			Use of Wien's law attempted [even with slips] [1]						
			Luminosity (or $L / L_\odot$) **and** peak wavelength of either evolved stage calculated [1.3×10^{30} W / $3000 L_\odot$ 970 nm; 5.6×10^{24} W / $1.5 \times 10^{-2} L_\odot$ 150 nm] [1]						
			[or luminosity or peak wavelength of both]						
			Characteristics of both stages calculated. [1]						
			Identification of colours [red $\longrightarrow$ blue/white] or region of e-m spectrum of peaks [IR, UV] [1]						
Total				3	5		8	4	

Rhodri's answers

(a) A black body absorbs and emits all radiation which falls on it. ✓ ✗

MARKER NOTE

Some confusion in Rhodri's answer. A black body absorbs all radiation which is incident upon it. Emitting 'all radiation which falls on it' makes no sense. It emits radiation more strongly at any wavelength than a non-black body. **1 mark**

(b) This is the total power output ✗ Not enough

MARKER NOTE

Rhodri was almost there but he needed to specify electromagnetic radiation. Stars also give out energy in the form of neutrinos, which does not contribute to the luminosity. **0 marks**

(c) The next stage is a Red giant – it would appear red and much brighter (more luminosity). The last stage is a White dwarf and much fainter. ✓

Red giant: $L = A\sigma T^4$

$= 4\pi (3.0 \times 10^8)^2 \times 5.67 \times 10^{-8} \times 3000^4$ ✓ attempt

$= 5.19 \times 10^{24}$ W

$\lambda_{max} = \dfrac{2.90 \times 10^{-3}}{3000} = 9.67 \times 10^{-7}$ – so red ✓

White dwarf: $L = A\sigma T^4$

$= 4\pi (1.4 \times 10^4)^2 \times 5.67 \times 10^{-8} \times 20000^4$ ✗

$= 2.2 \times 10^{19}$ W – much fainter

$\lambda_{max} = \dfrac{2.90 \times 10^{-3}}{20000} = 1.45 \times 10^{-7}$ – so white ✓

MARKER NOTE

The first mark that Rhodri achieves is the last one in the mark scheme: he correctly identifies the colours of the star in the two stages.

He picks up one mark each for using the Stefan-Boltzmann law and the Wien displacement law. He makes the same mistakes in the calculation of L for both stars: he fails to convert km to m and he uses the diameter instead of the radius in calculating the surface area. However, he correctly calculates the peak wavelength of both so he gains the 4th mark. The absence of the unit for λ_{max} is not further penalised. **4 marks**

Total **5 marks / 8**

Ffion's answers

(a) It absorbs all radiation which falls on it. ✓

It emits all wavelengths of radiation better than any other kind of body ✓

MARKER NOTE

The first part of the answer is good. Ideally, the second part of the answer would have mentioned that the comparison object was at the same temperature, e.g. a black body at 1000 K emits more radiation at all wavelengths than does a non-black body at 1000 K. However, this omission was not penalised on this occasion. **2 marks**

(b) Luminosity is the energy of the electromagnetic radiation given out every second. ✓

MARKER NOTE

A short and entirely correct answer. A well-learned definition. **1 mark**

(c) Main sequence → Red giant → White dwarf ✓

Luminosity: First stage,

$L = A\sigma T^4 = 4\pi(1.5 \times 10^{11})^2 \times 5.67 \times 10^{-8} \times (3000)^4$ ✓

$= 1.30 \times 10^{30}$ W $= 3400 \times L$ for Sun

Luminosity: Second stage

$L = A\sigma T^4 = \pi(1.4 \times 10^7)^2 \times 5.67 \times 10^{-8} \times (20000)^4$

$= 5.58 \times 10^{24}$ W $= 0.015 \times L$ for Sun ✓

Colour: Red giant

Peak $\lambda = \dfrac{W}{T} = \dfrac{2.90 \times 10^{-3}}{3000} = 9.7 \times 10^{-7}$ m ✓

This is in the infra-red region so the star will look red (and bright) as expected.

Colour: White dwarf

Peak $\lambda = \dfrac{W}{T} = \dfrac{2.90 \times 10^{-3}}{20000} = 1.5 \times 10^{-7}$ m ✓

This is in the UV so the star will appear bluish white and faint as expected.

MARKER NOTE

Ffion also achieves the 5th mark from the first part of her answer – a recollection of GCSE learning! Really, this answer should come out of the calculations but it scores the marks anyway.

Ffion uses both the Stefan-Boltzmann and Wien's laws. It appears she has made a slip in calculation of the luminosity of the white dwarf: she uses πd^2 instead of $4\pi r^2$. These, of course, give the same answer and the examiner awards the mark bod.

Her Wien calculations are correct. Hence she scores all 4 calculation marks, and her final written descriptions would have earned her the 5th mark if she hadn't already achieved it! **5 marks**

Total **8 marks / 8**

Component 2 Practice questions

Section 8: Orbits and the wider universe

Topic summary

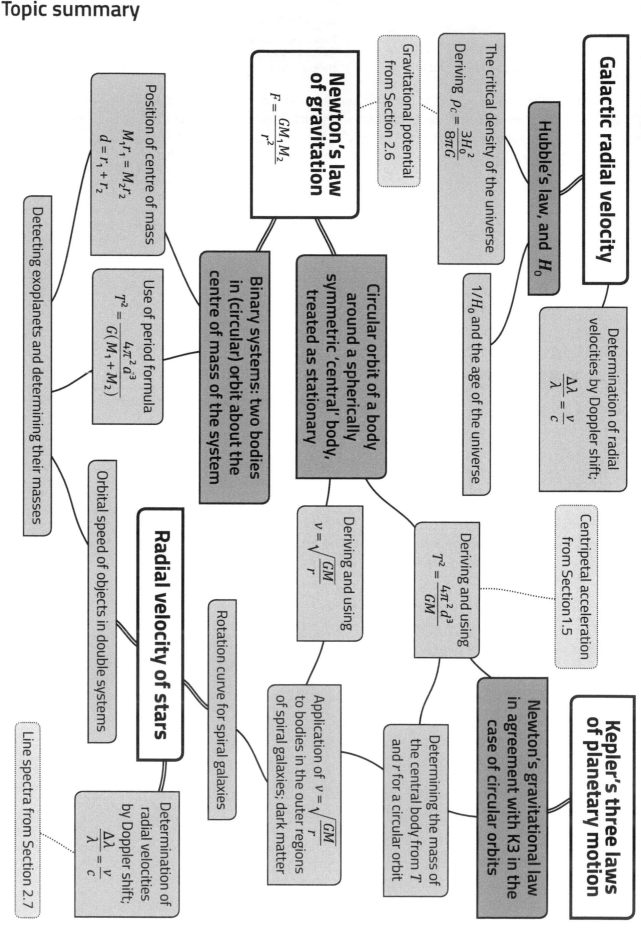

Galactic radial velocity

Hubble's law, and H_0

The critical density of the universe

Deriving $\rho_c = \dfrac{3H_0^2}{8\pi G}$

Gravitational potential from Section 2.6

$1/H_0$ and the age of the universe

Determination of radial velocities by Doppler shift; $\dfrac{\Delta\lambda}{\lambda} = \dfrac{v}{c}$

Newton's law of gravitation

$F = \dfrac{GM_1 M_2}{r^2}$

Position of centre of mass

$M_1 r_1 = M_2 r_2$

$d = r_1 + r_2$

Use of period formula

$T^2 = \dfrac{4\pi^2 d^3}{G(M_1 + M_2)}$

Detecting exoplanets and determining their masses

Binary systems: two bodies in (circular) orbit about the centre of mass of the system

Circular orbit of a body around a spherically symmetric 'central' body, treated as stationary

Orbital speed of objects in double systems

Radial velocity of stars

$v = \sqrt{\dfrac{GM}{r}}$

Deriving and using

$T^2 = \dfrac{4\pi^2 d^3}{GM}$

Deriving and using

Centripetal acceleration from Section 1.5

Rotation curve for spiral galaxies

Application of $v = \sqrt{\dfrac{GM}{r}}$ to bodies in the outer regions of spiral galaxies; dark matter

Determining the mass of the central body from T and r for a circular orbit

Newton's gravitational law in agreement with K3 in the case of circular orbits

Kepler's three laws of planetary motion

Determination of radial velocities by Doppler shift; $\dfrac{\Delta\lambda}{\lambda} = \dfrac{v}{c}$

Line spectra from Section 2.7

Q1 The Earth's orbit of the Sun is almost circular, with a radius of 150×10^6 km. Calculate a value for $M_\odot$, the mass of the Sun. [3]

Q2 (a) Assuming that the Earth is a sphere of radius 6370 km, show clearly that its mass is approximately 6×10^{24} kg. Use standard value for G and for g at the Earth's surface. [3]

(b) For your calculation in (a) to be valid, what do we have to assume about the way the Earth's mass is distributed? [1]

(c) Using your answer to (a), calculate a value for the mean density, ρ_{Earth}, of the Earth. [2]

Q3 A student is given a sketch of a comet's orbit around a star, and is asked to mark the positions of the star (S) and the point (X) in its orbit at which the planet moves most slowly. Here is his attempt:

Discuss whether the student's positions for S and X could be correct. [3]

Q4 (a) Astronomers commonly use the *astronomical unit* (AU) as a unit of distance in the solar system. This is the mean distance from the Earth to the Sun. Show that 1 AU is approximately 150 million km. [Mass of Sun = 1.99×10^{30} kg.] [2]

....

....

....

(b) Show that the light year (the distance travelled by light in one earth year) is approximately 9.5×10^{12} km. [2]

....

....

....

Q5 (a) The Moon orbits the Earth in an almost circular orbit of radius 3.83×10^5 km. Its period of revolution is 27.3 days. Hence show that the pull of the Earth must provide it with a centripetal acceleration of approximately 3×10^{-3} m s^{-2}. [3]

....

....

....

....

....

(b) (i) Hence calculate the ratio:

$$\frac{\text{acceleration due to Earth's gravity at Earth's surface}}{\text{acceleration due to Earth's gravity at Moon's distance}}$$ [1]

....

....

....

(ii) Treating the Earth as a sphere of radius 6.37×10^6 m, calculate the ratio:

$$\left(\frac{\text{distance of Moon from Earth's centre}}{\text{distance of Earth's surface from Earth's centre}}\right)^2$$ [1]

....

....

....

(c) Explain how these results support Newton's law of gravitation. [Newton did the equivalent calculations with the data available in his day.] [2]

....

....

....

....

Q6 (a) Calculate the height above the Earth's surface at which a geostationary satellite must orbit.
[Earth's radius = 6.37×10^6 m, Earth's mass = 5.97×10^{24} kg.] [4]

(b) State another requirement of the orbit for the satellite to be geostationary. [1]

Q7 Here are data for the (nearly circular) orbits of the two moons of Mars.

	Radius / 10^6 m	Period / day
Deimos	23.46	1.263
Phobos	9.39	0.319

Evaluate whether or not Kepler's 3rd law applies to these moons. [3]

Q8 An equation for the critical density of a 'flat' universe is:

$$\rho_c = \frac{3H_0^2}{8\pi G}$$

(a) Explain what is meant by *critical density*. [1]

(b) Show that the equation is correct as far as dimensions (or SI units) are concerned. [1]

Q9 The wavelength of an absorption line in a star's spectrum is found to vary regularly between 393.14 nm and 393.82 nm. Its wavelength as measured in the laboratory is 393.36 nm.

(a) Calculate the extreme values of the radial velocity of the star. [3]

....

....

....

....

(b) Without further calculation, describe a likely way in which the star is moving. Give reasons for your answer. [3]

....

....

....

....

Q10 Two stars, S_1, of mass 1.5×10^{30} kg, and S_2, of mass 2.5×10^{30} kg, travel in circular orbits about their common centre of mass, C. The stars are separated by 3.0×10^{12} m.

Calculate:

(a) The periodic time. [3]

....

....

....

....

....

(b) The radii of the orbits. [2]

....

....

....

....

Q11   Radial velocity is plotted against time for the star Tau Boötis A.

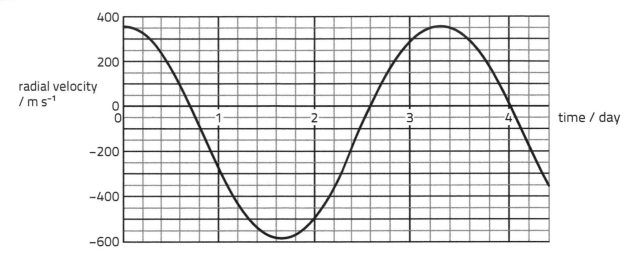

Assuming that its orbit is circular and seen edge-on:

(a) Show that the orbital velocity is approximately 500 m s⁻¹.   [2]

(b) Calculate the radius of the orbit.   [2]

(c) The mass of Tau Boötis A is estimated to be 2.6 × 10³⁰ kg. Its orbital motion is due to an unseen planet of much smaller mass. Calculate an approximate value for the radius of the planet's orbit.   [3]

(d) Calculate a value for the planet's mass.   [2]

Q12 The mass, m_{vis}, of a visible star is estimated as 12×10^{30} kg. The variation from the mean of its radial velocity is plotted against time.

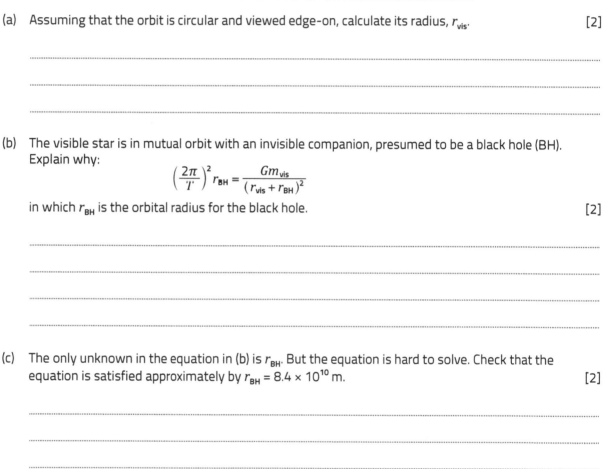

(a) Assuming that the orbit is circular and viewed edge-on, calculate its radius, r_{vis}. [2]

....

....

....

(b) The visible star is in mutual orbit with an invisible companion, presumed to be a black hole (BH). Explain why:

$$\left(\frac{2\pi}{T}\right)^2 r_{BH} = \frac{Gm_{vis}}{(r_{vis} + r_{BH})^2}$$

in which r_{BH} is the orbital radius for the black hole. [2]

....

....

....

....

(c) The only unknown in the equation in (b) is r_{BH}. But the equation is hard to solve. Check that the equation is satisfied approximately by $r_{BH} = 8.4 \times 10^{10}$ m. [2]

....

....

....

....

(d) Calculate the mass of the black hole. [2]

....

....

....

....

Q13 The diagram plots the radial velocities of galaxies against their distance away. The distances are measured in megaparsec (Mpc). 1 Mpc = 3.09 × 10²² m.

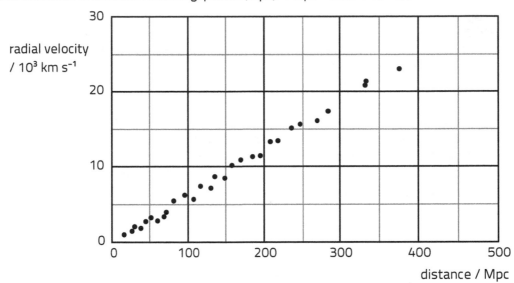

(a) Calculate a value for the Hubble constant, H_0, in SI units. [4]

(b) Give one reason, not involving errors of measurement, for the scatter of points. [1]

Q14 In a binary system, the two stars are separated by 30 AU and orbit the system's centre of mass in circular orbits of period 82.2 year. The orbital radius of the more massive star is 7.5 AU. Calculate the masses of the individual stars in terms of the solar mass, $M_\odot$.

[1 AU = mean distance from the Earth to the Sun = 1.50 × 10¹¹ m; $M_\odot$ = 1.99 × 10³⁰ kg] [5]

Component 2 Practice questions

Question and mock answer analysis

Q&A 1

(a) A body of mass, m, is in circular orbit of radius r around a spherically symmetric body of mass, M. [$M \gg m$.] Starting from Newton's law of gravitation, show clearly that the orbital speed, v, of the body is given by: $v = \sqrt{\dfrac{GM}{r}}$ [2]

(b) The great majority of the stars in a galaxy, G, are contained in its 'central region', whose mass can be estimated from its total luminosity. This estimated mass has been used with the equation of part (a) to plot the **broken** curve on the graph. This shows how the speed of orbiting matter in the outer region of the galaxy is expected to depend on its distance, r, from the centre of the galaxy.

The **full** curve shows how the *observed* speed of orbiting matter depends on r.

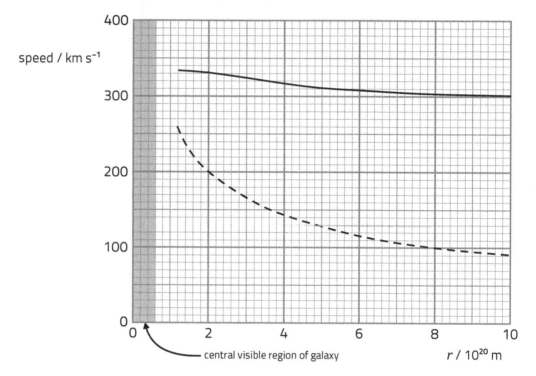

(i) Show that the **broken** curve is consistent with the equation in part (a). [2]

(ii) Determine the estimated mass of the galaxy, on which the **broken** curve is based. [3]

(iii) Explain what the two graphs suggest about the *actual* mass of the galaxy and the way in which the mass is distributed. [4]

(iv) Briefly describe one way that has been suggested to account for these apparent discrepancies regarding galactic mass. [1]

(c) Explain how the equation:

$$\frac{v}{c} = \frac{\Delta\lambda}{\lambda}$$

can be used to determine the speed of a body in a distant galaxy. [3]

What is being asked

(a) Here we have a quick and straightforward application of Newton's law of gravitation, in this piece of AO1 bookwork. 'Show clearly' means give some understandable working.

(b) (i) An AO3 question, with no hint as to how to proceed.

(ii) This tests that you can put data from the graph into the equation of part (a), rearranging as necessary. Two or three rather basic skills are needed, but it's easy to make mistakes in reading from a graph!

(iii) and (iv) These are AO1 questions. The implication of the observed variation of orbital speed of objects in the outer parts of a galaxy is expected in the specification. The difficulty here is in producing concise answers including an explanation of how they link to the data.

(c) Clearly you must state what the symbols in the equation mean, and which values are found by measurement. Do give details – but be aware of your limited time.

Mark scheme

Question part		Description	AOs			Total	Skills	
			1	2	3		M	P
(a)		$\dfrac{mv^2}{r} = \dfrac{GMm}{r^2}$ **or** $mr\omega^2 = \dfrac{GMm}{r^2}$ [1] Any correct intermediate algebraic step. [1]	2			2 1		
(b)	(i)	Two data points chosen, e.g. (2, 200), (8, 400) [1] Convincing manipulation $\longrightarrow v \propto r^{-0.5}$ [1]			2	2	1	
	(ii)	$M = \dfrac{rv^2}{G}$ (rearrangement at any stage) [1] Recognisable readings with correct powers of 10 taken from a point on the broken curve. [1] $M = 1.2$ or 1.3×10^{41} kg [1] Accept 3 sf: No *second* penalty for wrong powers of 10			3	3	1	
	(iii)	Actual mass greater than estimated [1] because speed higher [throughout] [1] Mass extends out [further than supposed] from centre. [1] Because speed hardly decreases with distance from centre, or equiv. [1]	4			4		
	(Iv)	**Either** Dark matter with some explanation, e.g. not emitting light (accept 'hidden') or throughout galaxy **or** Newton's law of gravitation fails. [1]	1			1		
(c)		Measure wavelength of light from the body [1] In the equation, $\Delta\lambda$ = shift in wavelength, λ = [expected] wavelength, v = speed of body. [1] One extra detail, e.g. [identifiable] emission **or** absorption line used, **or** galaxy needs to be edge-on, **or** positive $\Delta\lambda$ shows body moving away from us. [1]	3			3		
Total			10	3	2	15	3	

Rhodri's answers

(a) $\dfrac{GMm}{r^2} = mr\omega^2$ ✓

$\dfrac{GM}{r^2} = r\dfrac{v^2}{r}$ ✗

$v^2 = \dfrac{GM}{r^2}$

I don't know why this is wrong.

> **MARKER NOTE**
> Rhodri has chosen the formula for centripetal acceleration that is less convenient for the problem in hand. But his first equation is correct and scores the 1st mark. No more marks, though, because, in trying to get rid of ω, he confused $\omega = v/r$ with $a = v^2/r$.
>
> **1 mark**

(b) (i) When r doubles from $2 \rightarrow 4 \times 10^{20}$ m the speed drops from 200 to 140 km s^{-1} ✓ This is approximately half so there is inverse proportion. ✗

> **MARKER NOTE**
> Rhodri has hit the 1st marking point, in comparing two data points. He has not understood the significance of the square root for this answer and so misses the 2nd mark.
>
> **1 mark**

(ii) When $r = 2 \times 10^{20}$, $v = 200$ ✓

So $200^2 = \dfrac{6.67 \times 10^{-11}M}{2 \times 10^{20}}$ ✓

$M = \dfrac{200^2 \times 2 \times 10^{20}}{6.67 \times 10^{-11}} = 1.2 \times 10^{35}$ kg ✗

> **MARKER NOTE**
> Rhodri's answer is correct except for not taking account of the km s^{-1} unit on the vertical scale. This is a common type of error, but the rather generous mark scheme penalises it only once. It's slightly long-winded to make M the subject *after* putting in the numbers, but whatever you're comfortable with...
>
> **2 marks**

(iii) The graph suggests that the mass of the galaxy is greater than that estimated. ✓ This follows from the equation in part (a) because the actual speed is much faster than that using the estimated mass, ✓ and the shapes of the graphs are different. [not enough]

> **MARKER NOTE**
> Rhodri has understood that there is more than the estimated mass and has justified it well, by appealing to the graph and the equation. He has not developed his remark about the shapes of the graphs to say anything useful about the mass distribution.
>
> **2 marks**

(iv) One idea is that there is hidden mass called 'dark matter'. 'Hidden' means that it doesn't interact with e–m radiation, which is why we wouldn't be able to detect it. ✓ Dark matter may consist of little-understood particles such as WIMPs.

Another suggestion is that there isn't really any extra mass, but that Newton's law of gravitation doesn't always work.

> **MARKER NOTE**
> Rhodri is clearly well-informed about dark matter! His answer is almost too thorough for the 1 mark allocated, and there was certainly no need for two different suggested solutions. In this case, as fairly generally, there is no penalty for giving too much information.
>
> **1 mark**

(c) $\Delta\lambda$ is the Doppler shift ✓ [bod] of the light and λ is its wavelength. c is the speed of light. The equation gives the velocity v of the body, which is away from us if the wavelength is increased. ✓

> **MARKER NOTE**
> Rhodri lost the 1st mark because he didn't actually say where the light comes from! The 2nd mark is unsafe because 'Doppler shift' doesn't specify *wavelength* shift, but the examiner gave benefit of doubt because of the reference to increase in *wavelength* at the end. The last mark is gained for the 'extra detail' about the direction of the velocity.
>
> **2 marks**

| **Total** | **9 marks /15** |

Ffion's answers

(a) $\dfrac{mv^2}{r} = \dfrac{GMm}{r^2}$ ✓

$v^2 = \dfrac{GM}{r}$ (times by $\dfrac{r}{m}$) ✓

$v = \sqrt{\dfrac{GM}{r}}$

MARKER NOTE
Ffion's answer is clear and correct. She has even given a commentary on her algebra!

2 marks

(b) (i) If $v = \sqrt{\dfrac{GM}{r}}$, then $v^2 \propto \dfrac{1}{r}$ so $\left(\dfrac{v_1}{v_2}\right)^2 = \dfrac{r_2}{r_1}$.

Take $r_1 = 2$; $v_1 = 200$ and $r_2 = 8$; $v_2 = 100$ ✓

$(r_2/r_1) = 4$ and $(v_1/v_2)^2 = 2^2 = 4$. These are the same so OK. ✓

MARKER NOTE
Ffion homes in on two easily used data points, with one radius 4× the other for the 1st mark. Her clear algebra identifies a test and she uses it correctly for the 2nd mark.

2 marks

(ii) Rearranging the given equation

$M = \dfrac{rv^2}{G}$ ✓

$M = \dfrac{4 \times 10^{20} \times 180\,000^2}{6.67 \times 10^{-11}}$ ✓ $= 1.94 \times 10^{41}$ kg ✗

MARKER NOTE
Ffion's method is crystal clear, but she misread the vertical scale, leading to a seriously inaccurate answer. There was only one mark deducted, but a meaner mark scheme could have resulted in a 2-mark loss.

2 marks

(iii) At all values of r in the outer galaxy, the actual speed is greater than the calculated speed, ✓ so if the equation in part (a) is true, the actual mass must be greater than the estimated mass. ✓ Also, the speed hardly falls with increasing distance, as it would if almost all the mass were in the central region (v proportional to $1/\sqrt{r}$). ✓ So it looks as if there is much more mass than assumed in the outer regions. ✓

MARKER NOTE
Ffion has produced what amounts to a model answer. Both the larger than estimated mass and the wider distribution of mass are noted and justified from the graphs, and equation $v \propto \dfrac{1}{\sqrt{r}}$ was a nice touch!

4 marks

(iv) The galaxy contains material (particles) which we cannot detect with electromagnetic radiation, but which has mass. It is called 'dark matter'. ✓

MARKER NOTE
This is a clear and concise answer.

1 mark

(c) We examine light coming to us from the body. If we can recognise a wavelength as coming from a particular sort of atom ✓, and that wavelength is shifted by $\Delta\lambda$ from its usual value, λ, ✓ then the body is moving away from us with radial velocity $c \times \Delta\lambda/\lambda$, if $\Delta\lambda$ is positive. ✓

MARKER NOTE
Ffion's summary is very neat. She has explained the use of the equation and her reference to light coming from a particular sort of atom is a valuable extra detail.

3 marks

Total **14 marks /15**

Component 2 Practice questions

Practice papers

A LEVEL PHYSICS
COMPONENT 1 PRACTICE PAPER
[Section A only]

2 hours

For Examiner's use only		
Question	Maximum Mark	Mark Awarded
1.	9	
2.	13	
3.	10	
4.	10	
5.	12	
6.	9	
7.	6	
8.	11	
Total	80	

Notes

A live Eduqas Component 1 paper has two sections. The questions on the following pages represent a practice version of Section A. Section B consists of a passage of information on a physics topic followed by questions totalling 20 marks. You are recommended to practise for Section B by using past Eduqas and WJEC papers from the current and previous specifications.

In an Eduqas paper, the following information will be given on the front of the paper:

1. **Additional materials**
   You will be told that you will require a calculator and a **Data Booklet**. Sometimes you will be told that you need a ruler and/or an angle measurer / protractor.

2. **Answering the examination**
   You will be told to use a blue or black ball-point (but graphs are best drawn using a pencil).
   You will be told to answer **all** the questions in the spaces provided on the question paper.

3. **Further information**
   Each question part shows, using square brackets, the total marks available. One question will assess the quality of extended response [QER]. This question will be identified on the front page. In this practice paper the QER question is question **8**.

SECTION A

Answer **all** *questions.*

1. Andrea holds a uniform ladder of length 4.00 m and mass 22.4 kg as is shown in the diagram:

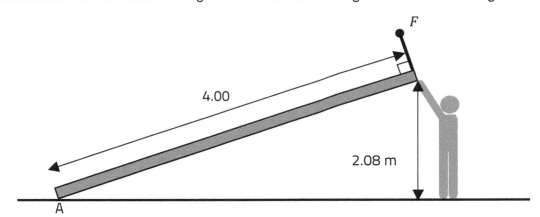

(a) Show that the force, F, that Andrea exerts on the ladder is approximately 100 N when the ladder is in equilibrium. [4]

(b) Calculate the force exerted by the ground on the ladder at point A and state its direction. [5]

2. Vivienne carries out an experiment to measure g, the acceleration due to gravity, by freefall. She uses the following set up:

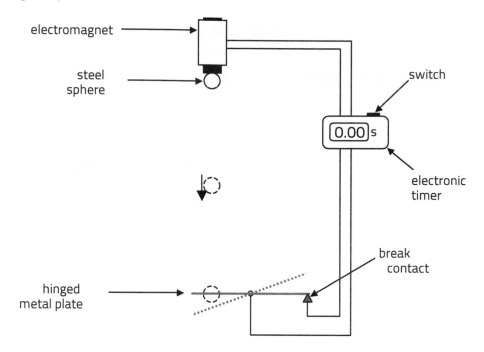

(a) Use the equation:

$$x = ut + \frac{1}{2}at^2$$

to show that a graph of time, t, against the square root of height ($\sqrt{h}$) should be a straight line through the origin with gradient $\sqrt{\frac{2}{g}}$. [2]

....

....

....

(b) Her results are recorded in the following table:

Height, h/m	$\sqrt{h}$/m	Freefall time / s				
		1	2	3	4	Mean, t
0.150	0.387	0.20	0.20	0.20	0.20	0.200
0.400	0.632	0.32	0.31	0.31	0.32	0.315
0.900	0.949	0.43	0.47	0.44	0.44	0.445
1.500	1.225	0.57	0.58	0.58	0.55	0.570
2.000	1.414	0.68	0.64	0.68	0.67	0.668

Table 1

(i) Plot a graph of t, the mean time, against $\sqrt{h}$, and draw a line of best fit. [3]

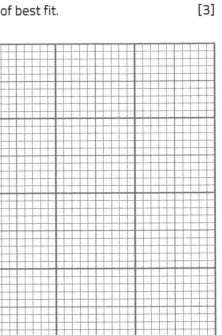

(ii) Use the best fit line to obtain a value of g. [3]

....

....

....

....

....

....

....

(iii) Explain to what extent the graph and gradient agree with the theory of part (a). [3]

....

....

....

....

....

....

(iv) Suggest a reason for a positive y-axis intercept in part (iii) and suggest how this could be corrected. [2]

....

....

....

3. Two snooker balls of mass 0.160 kg collide as is shown in the diagram:

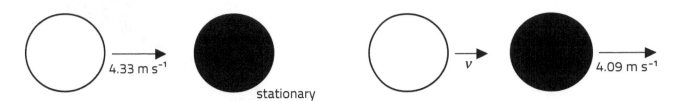

Before collision   After collision

(a) (i) State the principle of conservation of momentum. [2]

....

....

....

(ii) Calculate the speed, v, of the snooker ball shown after the collision. [2]

....

....

....

(iii) Determine how much energy is lost to the surroundings in this collision. [3]

....

....

....

....

(b) The snooker balls are in contact with each other for 16 μs during the collision.

Calculate the mean force acting on **each ball** during the collision. [3]

....

....

....

....

....

....

4. A snowboarder of total mass 89 kg descends a slope of length 360 m from rest and drops a vertical distance of 92 m.

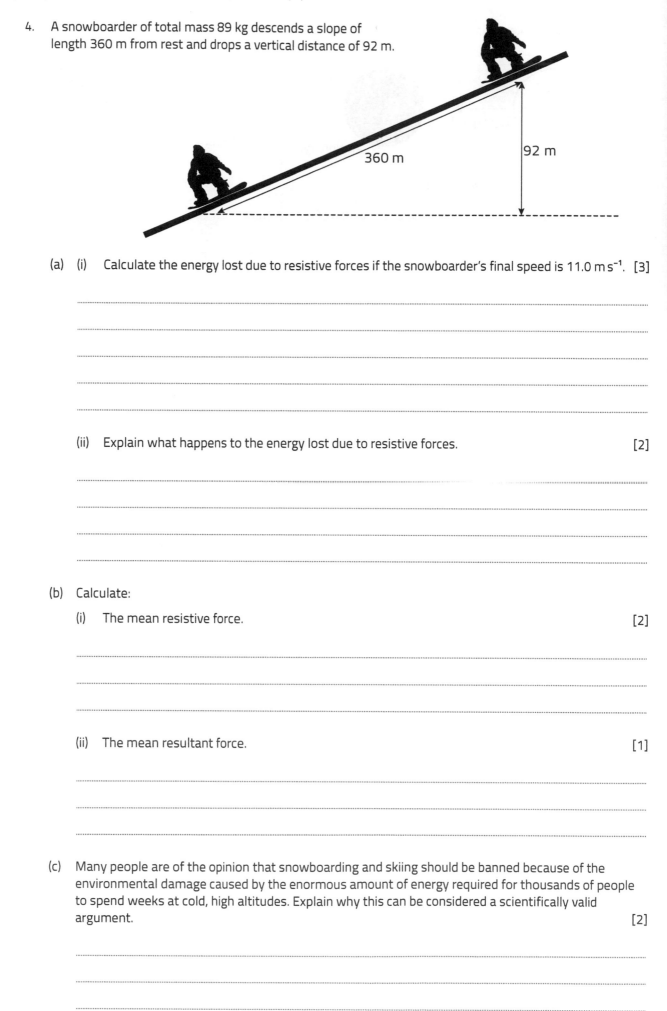

360 m

92 m

(a) (i) Calculate the energy lost due to resistive forces if the snowboarder's final speed is 11.0 m s⁻¹. [3]

....

....

....

....

....

(ii) Explain what happens to the energy lost due to resistive forces. [2]

....

....

....

....

(b) Calculate:

(i) The mean resistive force. [2]

....

....

....

(ii) The mean resultant force. [1]

....

....

....

(c) Many people are of the opinion that snowboarding and skiing should be banned because of the environmental damage caused by the enormous amount of energy required for thousands of people to spend weeks at cold, high altitudes. Explain why this can be considered a scientifically valid argument. [2]

....

....

....

5. A hemispherical bowl containing a small rubber ball is rotated about the vertical axis as shown. As the bowl is rotated, the rubber ball moves to an equilibrium position where it remains in contact with the same point on the surface of the bowl.

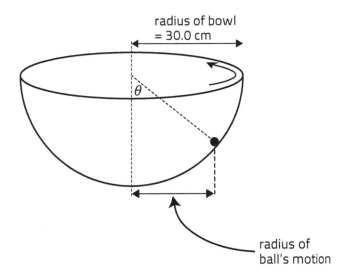

radius of bowl = 30.0 cm

θ

radius of ball's motion

(a) For the instantaneous position shown above, draw arrows to represent:

    (i)   The normal contact force acting on the rubber ball (label it R).   [1]

    (ii) The weight of the rubber ball (label it W).   [1]

(b) The centripetal force acting on the rubber ball is provided by the horizontal component of the normal contact force.

    (i)   Explain what is meant by the centripetal force.   [2]

    (ii) Explain how the following equation can be derived:

$$R\sin\theta = m\omega^2 r$$

    where R is the normal contact force, m is the mass of the rubber ball, ω is the angular velocity of the rubber ball (and bowl) and r is the radius of the motion of the ball.   [2]

    (iii) Explain how the following equation can be derived: [2]

$$R\cos\theta = mg$$

(iv) Jasmine states that when the radius of motion of the rubber ball is 15.0 cm the frequency of rotation of the bowl is approximately 1 Hz. Determine whether or not Jasmine is correct. [4]

....

....

....

....

....

....

....

6. A boy sits in a steel chair suspended by a spring as shown:

   (a) The boy has mass 45 kg and the spring constant of the spring is 5320 N m⁻¹. Calculate the distance the chair lowers when the boy sits in it. [3]

   ....

   ....

   ....

   ....

   (b) The boy's sister notices that the chair with her brother sitting in it performs SHM. She times five oscillations in 3.55 s. Determine whether her figures are consistent with part (a) and account for any discrepancy. [4]

   ....

   ....

   ....

   ....

   ....

   ....

   ....

   (c) State briefly the transfers of energy as the chair goes from the top of its motion to the bottom of its motion. [2]

   ....

   ....

   ....

7. An ideal gas is taken through the cycle ABCA shown below:

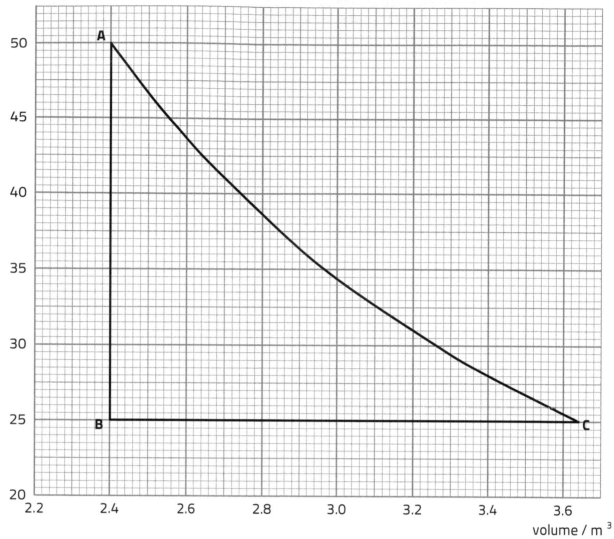

pressure / kPa

(a) The temperature at point A is 293 K. Calculate the number of moles of gas. [2]

....

....

....

(b) Calculate the temperatures at points B and C. [2]

....

....

....

(c) Calculate the work done by the gas for: [4]

AB

....

....

BC

....

....

CA

....

....

The cycle ABCA

....

....

(d) David states that no heat is transferred during CA. Determine whether or not David is correct. [3]

....

....

....

....

....

8. State the assumptions of the kinetic theory of gases and explain how these assumptions lead to gas molecules exerting pressure on their container (given by the equation $p = \frac{1}{3}\rho \overline{c^2}$). No mathematics is required.

[6 QER]

....

....

....

....

....

....

....

....

....

....

....

....

....

....

END OF PAPER

A LEVEL PHYSICS
COMPONENT 2 PRACTICE PAPER

2 hours

For Examiner's use only		
Question	Maximum Mark	Mark Awarded
1.	11	
2.	16	
3.	11	
4.	15	
5.	9	
6.	13	
7.	12	
8.	13	
Total	100	

Notes

In an Eduqas paper, the following information will be given on the front of the paper:

1. **Additional materials**
   You will be told that you will require a calculator and a **Data Booklet**. Sometimes you will be told that you need a ruler and/or an angle measurer / protractor.

2. **Answering the examination**
   You will be told to use a blue or black ball-point (but graphs are best drawn using a pencil).
   You will be told to answer **all** the questions in the spaces provided on the question paper.

3. **Further information**
   Each question part shows, using square brackets, the total marks available. One question will assess the quality of extended response [QER]. This question will be identified on the front page. In this practice paper the QER question is question **1**.

Component 2 Practice paper

Answer **all** questions.

1. (a) A typical force–extension graph is sketched for a metal wire:

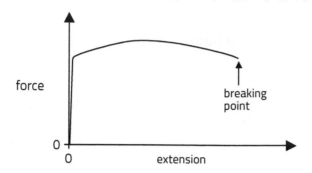

Explain, on the level of atoms, what is happening to the metal as it is stretched up to and including its breaking point, linking your explanation to the shape of the curve. [6 QER]

....

....

....

....

....

....

....

....

....

....

....

....

....

....

(b) Adam plans to hang a disco ball of mass 5.5 kg by steel wires of diameter 0.36 mm from a ceiling as shown:

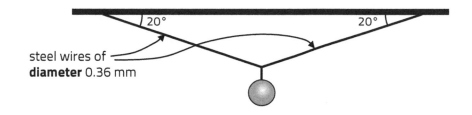

steel wires of
diameter 0.36 mm

20°   20°

Evaluate whether or not the wires are thick enough for their purpose.
[Breaking stress of the steel = 800 MPa.] [5]

2. (a) A potential difference is placed across two pieces of wire, made of the same metal alloy, in series (see diagram):

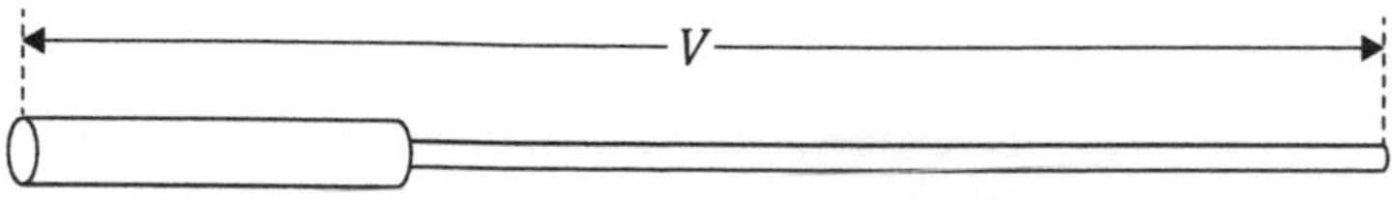

Wire X
Length L
Diameter d

Wire Y
Length $3L$
Diameter $d/3$

Showing your reasoning, determine these ratios:

(i) $\dfrac{\text{drift velocity of electrons in Y}}{\text{drift velocity of electrons in X}}$ [3]

(ii) $\dfrac{\text{power dissipated in Y}}{\text{power dissipated in X}}$ [2]

(b) Lucy investigated how the resistance, R, of a wire depends on its thickness. She used six pieces of wire made of the same metal alloy, connecting a digital multimeter (set to its *Ohms* range) across 2.500 m lengths of each in turn. She also measured the diameter, d, of each wire. Her plot of values of R, against $1/d^2$ is given on the next page.

(i) The diameters of the wires are all less than 0.5 mm. Suggest what instrument Lucy used to measure the diameter. [1]

(ii) State what precautions she should have taken to promote accuracy when measuring the diameter. [2]

(iii) Show clearly that the gradient of the graph is $\dfrac{4\rho L}{\pi}$ in which L is the length of a wire and ρ is the resistivity of the metal alloy. [2]

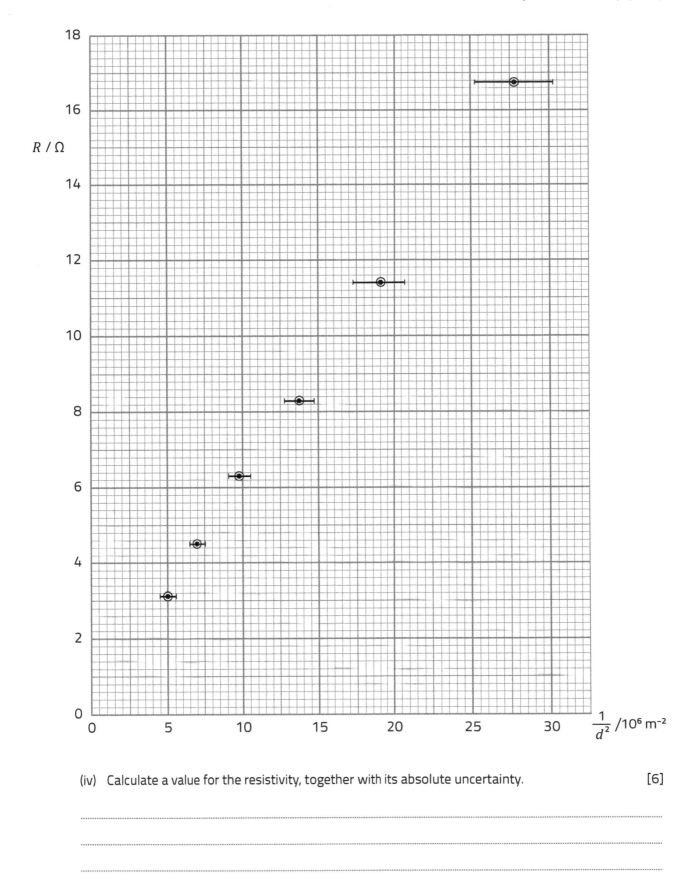

(iv) Calculate a value for the resistivity, together with its absolute uncertainty. [6]

....

....

....

....

....

....

3. (a) Define the *capacitance* of a capacitor. [1]

....

....

(b) An experimental capacitor consists of two metal plates, each measuring 0.25 m × 0.25 m, separated by air. It is found to have a capacitance of 3.00 nF.

(i) Show that the separation of the plates is roughly 0.2 mm. [2]

....

....

....

....

(ii) Calculate the *electric field strength* in the gap when the plates have charges of +60 nC and –60 nC. [3]

....

....

....

....

....

(iii) Suggest why a 3.00 nF capacitor may not be suitable for making a resistor-capacitor combination with a time constant of 30 s. Include a relevant calculation. [2]

....

....

....

....

(c) A 10 mF capacitor is charged to a pd of 12 V and then connected across a 15 mF capacitor, initially uncharged.

(i) Show that the new pd (across each capacitor) is approximately 5 V. [3]

....

....

....

....

....

(ii) The 10 mF capacitor originally stored 720 mJ of energy. Calculate the energy stored in the two capacitors together, after the connection is made.   [2]

....

....

....

....

(iii) Suggest what happens to the 'missing' energy.   [2]

....

....

....

4. (a) The diagram shows a potential divider circuit:

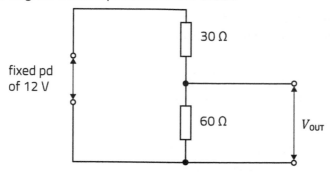

fixed pd of 12 V

30 Ω

60 Ω

V_{OUT}

(i) Evaluate whether a power rating of 1.0 W would be adequate for each resistor. [3]

....

....

....

....

....

(ii) Calculate the change in V_{out} when a load resistance of 180 W is connected across the output terminals. [5]

....

....

....

....

....

....

(b) A light-dependent resistor (LDR) has a resistance that decreases as the intensity of light falling on it increases. It is connected into the circuit shown:

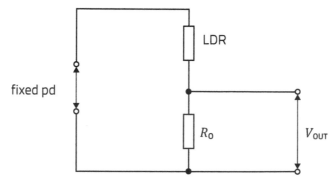

fixed pd

LDR

R_0

V_{OUT}

Paul claims that in daylight the output voltage, V_{out}, will be higher than in the dark, and that reducing the value of R_0 will increase V_{out}. Evaluate both these claims. [3]

....

....

....

....

5. The switch in the circuit shown is closed at time $t = 0$.

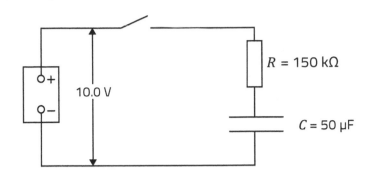

(a) On the axes provided sketch a graph to show how the charge, Q, on the capacitor plates varies with time. Scales are not required on the axes. **[2]**

(b) Calculate the time at which the charge reaches 50% of its final value. **[3]**

....

....

....

....

....

....

(c) The *current* in the circuit decreases as the capacitor becomes more fully charged.

(i) How can this be seen from the Q–t graph? **[1]**

....

(ii) Trisha says that you can understand why the current must decrease by considering the pd across the resistor and applying Ohm's law. Evaluate her claim. **[3]**

....

....

....

....

....

6. Two small spheres, each carrying a charge of 12.0 pC, are held 60 mm apart, as shown. O is the midpoint of the line joining the spheres. The perpendicular bisector of this line is labelled 'axis'.

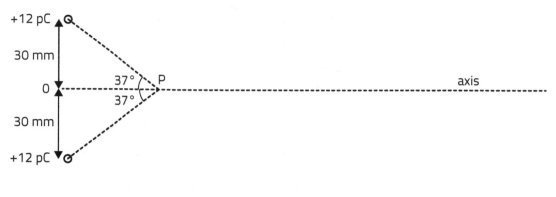

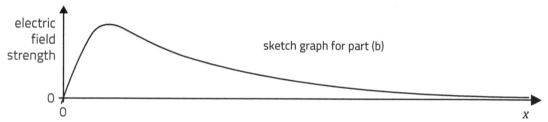

(a) Calculate the *resultant* electric field strength at point P, due to the spheres. [4]

....

....

....

....

....

....

....

(b) A student's sketch graph is given above. It attempts to show how the resultant electric field strength varies with distance x along the axis from O, to the right of O. Evaluate whether or not the shape of the graph could be correct. [3]

....

....

....

....

....

....

(c) (i) Define the *electric potential* at a point. [1]

....

....

(ii) A particle of mass 9.35×10^{-26} kg and charge -3.2×10^{-19} C leaves point O with a velocity of 5000 m s^{-1} to the right. Evaluate whether or not it will pass point P. [5]

....

....

....

....

....

....

....

....

7. The continuous spectrum of the star *Pollux* is given:

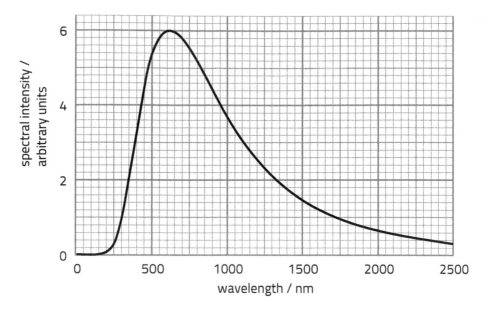

(a) Show that the temperature of Pollux is approximately 5000 K, stating the assumption on which your calculation is based. [3]

....

....

....

....

....

(b) The *luminosity* of Pollux is 1.65×10^{28} W.

(i) Explain briefly how this luminosity could have been determined from measurements made on Earth, knowing the distance, r_{PE}, of Pollux from the Earth. [2]

....

....

....

....

(ii) Determine the **diameter** of Pollux. [3]

....

....

....

....

....

....

(c)   The continuous spectrum of Pollux is crossed by many dark lines. Explain how these dark lines arise.

[4]

....

....

....

....

....

....

....

8. (a) Changes in a star's *radial velocity* as small as 0.3 m s⁻¹ can be detected by analysing the Doppler shift of light from the star. Calculate the wavelength shift corresponding to this change in radial velocity, when light of wavelength 397 nm is being observed. [2]

....

....

....

....

(b) The graph shows the variation in radial velocity of the star beta Geminorum (β Gem) over many Earth-days:

(i) Show that the **radius** of the (almost circular) path of β Gem is approximately 4×10^8 m.
[1 day = 86400 s.] [3]

....

....

....

....

....

(ii) The circular path of β Gem is due to a planet with a much smaller mass than the star itself. Making a suitable approximation, calculate the radius of the planet's orbit, taking the mass of β Gem to be 3.8×10^{30} kg. [3]

....

....

....

....

....

(iii) Calculate the mass of the planet. [2]

....

....

....

....

(c) Dean thinks that the only worthwhile astronomy using expensive facilities is the search for Earth-like *exoplanets* that might support life. Discuss whether or not this is a reasonable view to hold. [3]

....

....

....

....

....

....

END OF PAPER

Answers

Practice questions: Component 1: Newtonian Physics

Section 1: Basic physics

Q1
metre (m)   kilogram (kg)
second (s)   mole (mol)
ampère (A)   kelvin (K)

Q2 (a) newton (N)

(b) $N = kg\ m\ s^{-2}$

(c) Work = Force × distance moved in direction of force
So $J = N\ m = (kg\ m\ s^{-2})\ m = kg\ m^2\ s^{-2}$

Q3 (a) $k = \dfrac{F}{v}$, so $[k] = \dfrac{N}{m\ s^{-1}} = \dfrac{kg\ m\ s^{-2}}{m\ s^{-1}} = kg\ s^{-1}$

(b) $K = \dfrac{F}{Av^2}$, so $[K] = \dfrac{[F]}{[A][v^2]} = \dfrac{kg\ m\ s^{-2}}{m^2(m\ s^{-1})^2} = \dfrac{kg\ m\ s^{-2}}{m^4\ s^{-2}} = kg\ m^{-3}$

This is the unit of density, so K might represent the density of the fluid through which the object is moving, perhaps with a numerical (unit-less) multiplier.

Q4 $[\pi] = \dfrac{[A]}{[r^2]} = \dfrac{m^2}{m^2} = 1$, that is, no units.

Alternatively, we could show it using the equation: circumference = $2\pi r$

Q5 Scalars: energy, time, density, temperature, pressure
Vectors: acceleration, velocity, momentum
Note: pressure is a scalar because only the *magnitude* of a force is involved in its definition:

$$\text{Pressure} = \frac{\text{magnitude of force normal to a surface}}{\text{area of a surface}}$$

Q6 $[v] = m\ s^{-1}$; $[u] = m\ s^{-1}$; $[at] = m\ s^{-2} \times s = m\ s^{-1}$

So the two terms on the right-hand side have the same units, so can be added together **and** the unit of the left-hand side is the same as the unit of the right-hand side.

Q7 (a)   28 N   45 N

resultant magnitude 73 N   Direction = that of the 45 N force

(b)   45 N

        28 N
Resultant magnitude 17 N   Direction = that of the 45 N force

(c)   45 N

        28 N
53 N

Resultant magnitude = $\sqrt{45^2 + 28^2} = 53$ N   Direction = $\tan^{-1}\left(\dfrac{28\,N}{45\,N}\right) = 32°$ to 45 N force

Q8 (a) $F \sin 25° = 53$ so $F = \dfrac{53}{\sin 25°} = 125$ N **or** $F \cos(90° - 25) = 53$ leading to $F = 125$ N

(b) Horizontal component = $F \cos 25° = 125$ N $\times \cos 25° = 114$ N

Q9 (a) Perpendicular component = 5.0 kN × cos 75° = 1.3 kN [1.29 kN to 3 sf]

(b) 1.3 kN

(c) 1.29 kN = F × sin 10°

So $F = \dfrac{1.29 \text{ kN}}{\sin 10°}$ = 7.4 kN [7.43 kN to 3 sf]

(d) Since force components at right angles to forward direction cancel,

Resultant force = forward component of 5.0 kN + forward component of F
= 5.0 kN × cos 15° + 7.43 kN × cos 10°
= 4.83 kN + 7.32 N = 12 kN (to 2 sf)

Q10 (a) The vector sum of the forces on the object is zero.

(b) The sum of the clockwise moments about any point is equal to the sum of the anticlockwise moments about the same point.

Q11 (a) at t_1, v_{horiz} = 15.0 m s^{-1} × cos 30.0° = 13.0 m s^{-1}

v_{up} = 15.0 m s^{-1} × sin 30.0° (or cos 60.0°) = 7.5 m s^{-1}

(b) at t_2, v_{horiz} = 20.0 m s^{-1} × cos 49.5° = 13.0 m s^{-1}

v_{up} = −20.0 m s^{-1} × sin 49.5° (or cos 40.5°) = −15.2 m s^{-1}

So, Δv_{horiz} = 0

And Δv_{up} = (−15.2 m s^{-1}) − (+7.5 m s^{-1}) = −22.7 m s^{-1}

So the ball's change in velocity is 22.7 m s^{-1} in the downward direction.

Q12 Magnitude of $v_2 - v_1 = \sqrt{12^2 + 10^2}$ = 15.6 m s^{-1}

$\theta = \tan^{-1}\left(\dfrac{10}{12}\right)$ = 39.8°

∴ Bearing = 230° to nearest degree

Q13 $\left[\dfrac{k}{m}\right] = \dfrac{\text{N m}^{-1}}{\text{kg}} = \dfrac{(\text{kg m s}^{-2})\text{m}^{-1}}{\text{kg}} = \text{s}^{-2}$, so $\left[\dfrac{m}{k}\right] = \text{s}^2$

Now, $[T]$ = s, so the first two of the given equations are clearly wrong.

And $[T^2]$ = s^2, so the third equation could be right, but the fourth is wrong.

Q14 (a) $A = \pi r^2 = \pi \times \left(\dfrac{0.317 \times 10^{-3} \text{ m}}{2}\right)^2$ = 7.89 × 10^{-8} m^2

(b) Volume = Al = 7.89 × 10^{-8} m^2 × 0.85 m = 6.71 × 10^{-8} m^3

(c) Volume = $\dfrac{\text{mass}}{\text{density}} = \dfrac{2.50 \text{ kg}}{8.96 \times 10^3 \text{kg m}^{-3}}$ = 2.790 × 10^{-4} m^3

$l = \dfrac{\text{volume}}{A} = \dfrac{2.790 \times 10^{-4} \text{m}^3}{7.893 \times 10^{-8} \text{m}^2}$ = 3.54 × 10^3 m = 3.54 km

Q15 (a) Pressure = $\dfrac{\text{(magnitude of) force normal to a surface}}{\text{area of surface}}$

So (magnitude of) force = pressure × area = 2.5 × 10^6 Pa × π (0.100 m)^2 = 7.9 × 10^4 N to 2 sf

(b) Upward force of gas on piston = weight of copper piston + downward force of air on piston.

So $pA = Al\rho g + p_A A$
in which A is the cross-sectional area of the copper piston, l is its length and ρ is the density of copper, so Al is the volume of the copper cylinder, and $Al\rho$ is its mass.

Dividing through by A, and then substituting numerical data

$p = l\rho g + p_A$
= 0.10 m × 8.96 × 10^3 kg m^{-3} × 9.81 m s^{-2} + 101 × 10^3 Pa
= 110 × 10^3 Pa

Q16 (a) Without the 4.17 cm width measurement, the range of width measurements is (4.28 – 4.24) cm = 0.04 cm; with the 4.17 cm it is 0.11 cm, so the 4.17 cm is well outside the range established by the other readings, and it is wise to discard it.

(b) The absolute uncertainty in the mass may be taken as 0.1 g, so its percentage uncertainty is (0.1 / 600) × 100 = 0.017 %. This is far lower than the percentage uncertainties in the lengths, widths and heights (see below) so can be ignored.

(c) Mean length, $l = \dfrac{6.35 + 6.38 + 6.34 + 6.38 + 6.37}{5} = 6.36$ cm

$\Delta l = \dfrac{6.38 - 6.34}{2}$ cm = 0.02 cm; so $p(l) = \dfrac{0.02}{6.36} \times 100\% = 0.31\%$

Mean width, $w = \dfrac{4.26 + 4.24 + 4.28 + 4.25}{4}$ cm = 4.26 cm

$\Delta w = \dfrac{4.28 - 4.24}{2}$ cm = 0.02 cm; so $p(w) = \dfrac{0.02}{4.26} \times 100\% = 0.47\%$

Mean height, $h = \dfrac{2.79 + 2.81 + 2.83 + 2.80 + 2.81}{5}$ cm = 2.81 cm

$\Delta h = \dfrac{2.83 - 2.79}{2}$ cm = 0.02 cm; so $p(h) = \dfrac{0.02}{2.81} \times 100\% = 0.71\%$

$\rho = \dfrac{m}{V} = \dfrac{599.5 \text{ g}}{6.36 \text{ cm} \times 4.26 \text{ cm} \times 2.81 \text{cm}} = 7.87$ g cm^{-3}

$p(\rho) = 0.31 + 0.47 + 0.71 = 1.5\%$
So $\Delta \rho = 7.87$ g cm^{-3} × ±0.015 = ±0.12 g cm^{-3},
i.e. $\rho = (7.87 \pm 0.12)$ g cm^{-3} or (7.9 ± 0.1) g cm^{-3}
[N.B. It's fine to work in kg and m, yielding $(7.87 \pm 0.12) \times 10^3$ kg m^{-3}.]

Q17 CoG of ruler is at 25.0 cm. If mass of ruler = m_R, using PoM (moments about the pencil):
0.100 kg × g × 0.140 m = $m_R g$ × 0.100 m
∴ m_R = 0.140 kg
Now, with the metal piece mass, m_m, moments about pencil again
$m_m g$ × 0.115 m = 0.140 kg × g × 0.125 m
∴ m_m = 0.152 kg

Q18 (a) The tensions in A and B are equal by symmetry: the links are the same distances from the edges of the sign, which is uniform. Each tension is equal to half the pull of gravity on the sign.
So tension = $\frac{1}{2}$ × 3.5 kg × 9.81 N kg^{-1} = 17.2 N

(b) A is 0.15 m from H, distance AB is 0.60 m, so B is 0.75 m from H. Centre of mass of bar is 0.45 m from H. Weight of bar = 1.5 kg × 9.81 N kg^{-1} = 14.7 N
Sum of clockwise moments = 17.2 N × 0.15 m + 17.2 N × 0.75 m + 14.7 N × 0.45 m
$\qquad\qquad\qquad\qquad\qquad = 22.1$ N m

(c) Sum of anticlockwise moments about H = sum of clockwise moments about H
So $T \cos (90° – 30°)$ × 0.75 m = 22.1 N m
So $T = \dfrac{22.1 \text{ N m}}{0.75 \text{ m } \cos 60°} = 59$ N

(d) F_{right}, the horizontal force component to the right of the hinge on the bar, must balance the horizontal component of the wire's force on the bar.
∴ $F_{right} = T \cos 30° = 59$ N × cos 30° = 51 N
If F_{up} is upward force component of hinge on bar, then
F_{up} = weight of bar + sum of tensions in A and B – upward component of T
$\qquad = 14.7$ N + 34.3 N – 59 N × cos (90° – 30°) = 19.5 N
∴ $F = \sqrt{51^2 + 19.5^2} = 55$ N at $\tan^{-1}\left(\dfrac{19.5}{51}\right) = 21°$ above the horizontal to the right

Q19 (a) Work is a scalar not a vector; displacement is a vector not a scalar.

(b) (i) $[\mu] = [\rho][u][L] = \text{kg m}^{-3} \times \text{m s}^{-1} \times \text{m} = \text{kg m}^{-1}\text{s}^{-1}$
$\text{N s m}^{-2} = (\text{kg m s}^{-2}) \times \text{s} \times \text{m}^{-2} = \text{kg m}^{-1}\text{s}^{-1}$
So the units of the right- and left-hand sides of the equation are the same.

(ii) $k = \dfrac{\rho u L}{\mu} = \dfrac{1.16 \text{ kg m}^{-3} \times 41.2 \text{ m s}^{-1} \times 0.071\text{m}}{1.87 \times 10^{-5} \text{ N s m}^{-1}} = 1.8 \times 10^5$ (2 sf) / 1.81×10^5 (3 sf)

Section 1.2: Kinematics

Q1 (a) (i) Mean speed $= \dfrac{\text{total distance travelled}}{\text{total time taken}}$

(ii) Mean velocity $= \dfrac{\text{total displacement}}{\text{total time taken}}$

(b) (i) Mean speed $= \dfrac{120\,\text{m} + 120\,\text{m}}{27\,\text{s}} = \dfrac{240\,\text{m}}{27\,\text{s}} = 8.9 \text{ m s}^{-1}$ (2 sf)

(ii) Total displacement $= 120\sqrt{2}$ m on a bearing of 045° (in a NE direction)

∴ Mean velocity $= \dfrac{120\sqrt{2}}{27} \text{ m s}^{-1} = 6.3 \text{ m s}^{-1}$ on a bearing of 045° (NE)

Q2 Taking east as the positive direction:
Initial velocity, $u = 19 \text{ m s}^{-1}$; final velocity $v = -11 \text{ m s}^{-1}$
$\Delta v = v - u = -11 - 19 = -30 \text{ m s}^{-1}$; time, $t = 25$ ms $= 0.025$ s

∴ Mean acceleration $= \dfrac{\Delta v}{t} = \dfrac{-30 \text{ m s}^{-1}}{0.025 \text{ s}}$ eastwards $= 1200 \text{ m s}^{-2}$ westwards

This is a very large acceleration (about 120g) because the duration is so small.

Q3 (a) Using the labelled graph

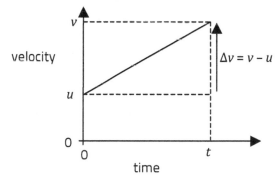

(i) Acceleration, $a = \dfrac{\Delta v}{t} = \dfrac{v - u}{t}$
Multiplying by t: $at = v - u$
Rearranging gives $v = u + at$ as required.

(ii) Displacement, x = area under the v–t graph = mean height × base
$= \dfrac{u + v}{2}t$

(iii) Displacement, x = area under the v–t graph
$\qquad$ = area of rectangle + area of triangle
$\qquad = ut + \dfrac{1}{2}(v - u)t$
But $at = v - u$ so $x = ut + \dfrac{1}{2}at^2$

(b) Rearranging $v = u + at$ gives $t = \dfrac{v - u}{a}$.
Substitute for t in $x = \dfrac{u + v}{2}t \longrightarrow x = \dfrac{(u + v)(v - u)}{2a}t$
Multiply by $2a$ and expand the brackets $\longrightarrow 2ax = v^2 - u^2$. Hence $v^2 = u^2 + 2ax$.

Q4 (a) Taking upwards as positive

    (i) $u = 15.5$ m s^{-1}, $a = -g = -9.81$ m s^{-2}; $v = 0$ (at the top point); $x = h$ (max height)

        $v^2 = u^2 + 2ax$, $\therefore 0 = (15.5)^2 - 2 \times 9.81h$

        $\therefore$ max height $h = \dfrac{(15.5)^2}{2 \times 9.81} = 12.2$ m

    Alternatively: use conservation of energy. $mgh = \frac{1}{2}mv^2$ and divide by m.

    (ii) $v = u + at$, so $t = \dfrac{v - u}{a} = \dfrac{0 - 15.5}{-9.81} = 1.58$ s

(b) The ball is decelerating, so the mean velocity in the first half of the ascent is greater than in the second. Hence the time taken to reach half the maximum height is less than half the time to reach the maximum.

(c) (i) Calculate time to drop 6.1 m from highest point.

        Take downwards as positive: $u = 0$, $a = g = 9.81$ m s^{-2}, $x = 6.1$ m.

        $x = ut + \frac{1}{2}at^2$, so $6.1 = \frac{1}{2} \times 9.81\,t^2$.

        So $t = \sqrt{\dfrac{2 \times 6.1}{9.81}} = 1.12$ s.

        $\therefore$ Total time $= 1.58$ s $+ 1.12$ s $= 2.70$ s

        Alternatively: calculate time from beginning with: $u = 15.5$ m s^{-1}, $a = -g$, $x = 6.1$ m

        Use the same equation: $x = ut + \frac{1}{2}at^2$ which gives $6.1 = 15.5t - \frac{1}{2} \times 9.81\,t^2$

        Solving the quadratic for $t \longrightarrow 0.46$ s and 2.70 s. Choose the second solution.

    (ii) Calculate the speed for the 6.1 m drop from the highest point:

        Time to drop this far $= 1.12$ s, from (c)(i), taking downwards as positive

        $v = u + at$, so $v = 0 + 9.81 \times 1.12 = 11.0$ m s^{-1}

        Alternatively: Use $v = u + at$ from the beginning

        With upwards positive $u = 15.5$ m s^{-1}, $a = -g$, $t = 2.70$ s

        So $v = 15.5 - 9.81 \times 2.70 = -11.0$ m s^{-1} hence a downward velocity of 11.0 m s^{-1}

        Or Calculate the speed gained in dropping 6.1 m using $v^2 = u^2 + 2ax$

Q5 (a) The water takes some time to reach the ground. When it is released it has a horizontal velocity. It keeps this horizontal motion when falling, so it needs to be released before getting to the drop zone.

(b) Let $t =$ time to fall 100 m. Use $x = ut + \frac{1}{2}at^2$ with $u = 0$, $a = g$ and $x = 100$ m.

    $\therefore t = \sqrt{\dfrac{2 \times 100}{9.81}} = 4.52$ s.

    In 4.52 s, the plane travels 120 m s^{-1} $\times$ 4.52 s $= 540$ m (2 sf), so water must be dropped 540 m before the burning area.

Q6 (a) (i) $u_h = u\cos\theta = 20.0 \times \cos 37° = 16.0$ m s^{-1}

    (ii) $u_v = u\sin\theta = 20.0 \times \sin 37° = 12.0$ m s^{-1}

    (iii) The resultant of 16 m s^{-1} horizontally and 12 m s^{-1} vertically is not $16 + 12 = 28$ m s^{-1} but $\sqrt{16^2 + 12^2} = 20$ m s^{-1}, because they are at right angles, so no worries!

(b) (i) Upwards is positive. $u_v = 12.0$ m s^{-1}, $a = -g$, $v_v = 0$, $x = h$ (height)

        $v^2 = u^2 + 2ax$, $\therefore 0 = (12.0)^2 - 2 \times 9.81h$

        $\therefore h = \dfrac{(12.0)^2}{2 \times 9.81} = 7.34$ m

    (ii) Let $t =$ time to reach ground. Consider vertical motion

        $x = ut + \frac{1}{2}at^2$, $\therefore 0 = 12.0t - 4.905t^2$, $\therefore t(12.0 - 4.905t) = 0$

        $\therefore t = 0$ (ignore) or $4.905t = 12 \longrightarrow t = 2.45$ s

        [Alternative method: twice the time to the maximum height.]

        $\therefore$ Horizontal distance travelled $= 16.0$ m s^{-1} $\times 2.45$ s $= 39.2$ m (3 sf)

(c) (i) The sin function is a ratio, so has no units.

        $\therefore \left[\dfrac{u^2\sin 2\theta}{g}\right] = \dfrac{(\text{m s}^{-1})^2}{\text{m s}^{-2}} = \text{m} = [R]$. So homogeneous.

(ii) Using the equation, $R = \dfrac{(20)^2 \sin 74°}{9.81} = 39.2$ m, i.e. the equation gives the same answer (at least to 3 sf).

Q7 (a) (i) From the graph, $v(20\text{ s}) = 7.95$ m s^{-1}; $v(10\text{ s}) = 3.70$ m s^{-1}

$\therefore$ Mean acceleration = $\dfrac{(7.95 - 3.70)\text{m s}^{-1}}{10.0\text{ s}}$ 0.43 m s^{-2} (to 2 sf)

(ii)

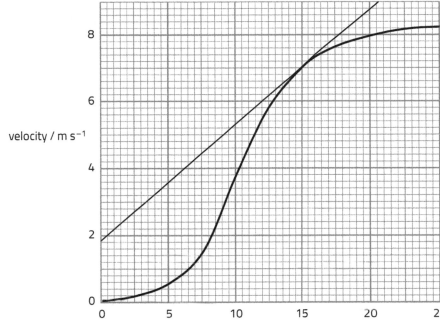

Acceleration at 15.0 s = gradient of tangent at 15.0 s

$= \dfrac{(9.0 - 1.8)\text{m s}^{-1}}{(20.7 - 0.0)\text{s}}$

$= 0.35$ m s^{-2}.

(b) The chord to the graph from 14.5 s to 15.5 s is almost indistinguishable from the tangent at 15.0 s, so its gradient is almost identical. Hence to a good approximation this method should work. [However it is impossible to carry this out accurately because reading such small distances on the graph has a large fractional uncertainty.]

Q8 (a) Total displacement = area under graph

$= \frac{1}{2} \times (15 + 29)\text{ s} \times 14\text{ m s}^{-1}$

$= 308$ m

$\therefore$ Mean velocity $= \dfrac{308\text{ m}}{29\text{ s}} = 10.6$ m s^{-1} (3 sf)

(b)

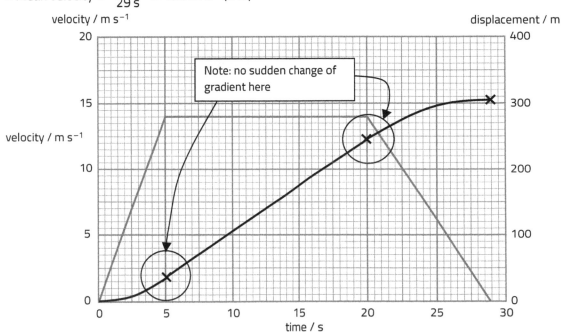

Note: no sudden change of gradient here

Calculations: Displacement at 5 s = 0.5 × 5 × 14 = 35 m
Displacement between 5 and 20 s = 15 × 14 = 210 m [⟶ total 245 m]
Displacement between 20 and 29 s = 0.5 × 9 × 14 = 63 m [⟶ total 308 m]

(c)

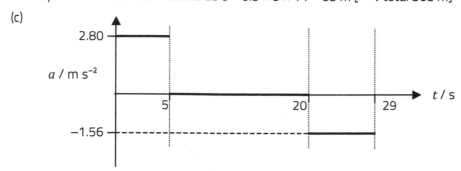

Q9 (a) Vertical motion; upwards positive.

Time ball in air = $\dfrac{v - u}{a} = \dfrac{(10.0 - (-10.0))\ \text{m s}^{-1}}{9.81\ \text{m s}^{-2}} = 2.04$ s.

∴ Distance train moves = $8.0\ \text{m s}^{-1} \times 2.04 = 16.3$ m

(b)

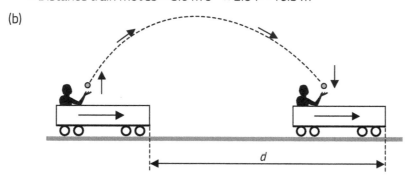

(c) When Helen throws the ball upwards, it also has a horizontal velocity. This horizontal velocity remains unchanged through the motion (ignoring air resistance) so when it comes down it has moved the same distance forward as Helen.

(d) From Helen's point of view, there is a backwards wind which pulls the ball behind her. From the observer's point of view, the ball is moving though the air, it experiences air resistance, which slows it down so it doesn't travel as far forward as Helen and lands behind her.

Section 1.3: Dynamics

Q1 (a) In an interaction between two bodies, A and B, the force exerted by body B on body A is equal in magnitude and opposite in direction to the force exerted by body A on body B.

(b) (i)

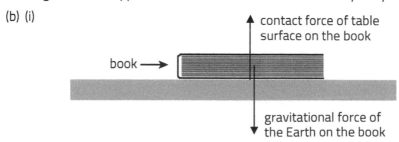

(ii) Contact force: N3 partner acts on the table surface
Gravitational force: N3 partner acts on the (whole) Earth

Q2 (a) Starting with Angharad's equation, the momentum, p, of a body is defined as mv (with the usual symbols). So for a constant mass:

$$F = \frac{\Delta p}{t} = \frac{\Delta(mv)}{t} = m\frac{\Delta v}{t}$$

Acceleration, a, is defined as $\dfrac{\Delta v}{t}$, so $F = ma$, which is Bethan's equation.

(b) There is not a single mass which is being accelerated, for which Bethan's equation would be useful. However, we can calculate the change in momentum of each molecule and therefore the total momentum change per second at the wall, so Angharad's is more useful.

Q3 (a) $v^2 = u^2 + 2ax \therefore a = \dfrac{v^2 - u^2}{2x} = \dfrac{(2.1 \text{ m s}^{-1})^2 - (1.5 \text{ m s}^{-1})^2}{2 \times 2.7 \text{ m}} = 0.40 \text{ m s}^{-2}$

Resultant force, $ma = 28 \text{ kg} \times 0.40 \text{ m s}^{-2} = 11.2 \text{ N}$

(b) If the frictional force on the box is F, the resultant force = 18.2 N – F
$\therefore$ 18.2 N – F = 11.2 N
$\therefore F$ = 18.2 N – 11.2 N = 7.0 N to the left
So the frictional force of box on ground = 7.0 N to the right

Q4 (a) The resultant force on the trolleys together = ma = 26 kg × 0.75 m s^{-2} = 19.5 N
The total frictional force = 10.0 N
$\therefore$ If F = force exerted by rope, F – 10.0 N = 19.5 N
$\therefore F$ = 29.5 N

(b) Resultant force on trolley B = 12 kg × 0.75 m s^{-2} = 9.0 N
Frictional force = 5.0 N
$\therefore$ force exerted by chain on trolley B = 9.0 N + 5.0 N = 14.0 N (forwards)
$\therefore$ By Newton's 3rd law, force exerted by B on chain = 14.0 N (backwards)

Alternatively: Resultant force on trolley **A** + chain = 14 kg × 0.75 m s^{-2} = 10.5 N
Frictional force = 5.0 N (backwards); force exerted by rope = 29.5 N (forwards)
$\therefore$ Backwards force exerted by **B** on chain = 29.5 N – 10.5 N – 5.0 N = 14.0 N

Q5 Magnitude of resultant force,

$R = \sqrt{8.0^2 + 16.0^2} = 17.9$ N (3 sf)

Direction: $\theta = \tan^{-1}\left(\dfrac{16.0}{8.0}\right) = 63.4°$

$\therefore$ Acceleration, $a = \dfrac{F}{m} = \dfrac{17.9 \text{ N}}{4.0 \text{ kg}} = 4.48 \text{ m s}^{-2}$

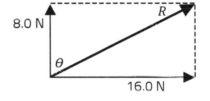

So the acceleration is 4.5 m s^{-2} on a bearing of 063°
(both to 2 sf)

Q6 (a) Vertical components of force cancel.
Resultant horizontal force = 2 × 6.0 cos 30° = 10.4 N to the left.
$\therefore$ Acceleration $\dfrac{F}{m} = \dfrac{10.4 \text{ N}}{0.050 \text{ kg}} = 210 \text{ m s}^{-2}$ to the left (2 sf)

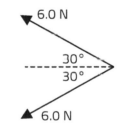

(b) Reason 1: the extension of the springs will decrease so the tension in them will also decrease.

Reason 2: The angle of the springs to the horizontal will increase so the horizontal component of the forces due to the tension will decrease.

Q7 (a) Let mass of each trolley = m.
Initial momentum to right = 8.0m – 2 × 2.2m = 3.6m
$\therefore$ By conservation of momentum, momentum after collision = 3.6m to the right

$\therefore$ Velocity of trolleys = $\dfrac{p}{\text{mass}} = \dfrac{3.6m}{3m} = 1.2 \text{ m s}^{-1}$ to the right.

(b) Initial kinetic energy = $\frac{1}{2}m \times (8.0)^2 + 2 \times \frac{1}{2}m \times (2.2)^2 = 36.84m$
Final kinetic energy = $3 \times \frac{1}{2}m \times (1.2)^2 = 2.16m$

Fraction of initial kinetic energy = $\dfrac{2.16}{36.84} = 0.059$

(c) A very small fraction of the lost energy will be given out as sound waves. The collision will produce compression waves in the metal frames of the trolleys which will temporarily lead to increased vibrations of the metal ions (atoms), hence this is likely to be true. This additional energy will quickly be transferred to the air by conduction and convection.

Q8   The total momentum (strictly 'linear momentum') of the dumbbell is zero because momentum is a vector and the two weights are moving at equal speeds in opposite directions.
The kinetic energy of each weight = $\frac{1}{2}mv^2$ = 1.25 × (5.0)^2 = 31.25 J

Kinetic energy is a scalar and has no direction.
∴ Total kinetic energy = 31.25 + 31.25 = 63 J (2 sf)

Q9   (a) Considering energy transfer for the falling ball,
$\frac{1}{2}mv^2 = mgh$ ∴ Impact speed = $\sqrt{2gh} = \sqrt{2 \times 9.81 \times 2.00}$ = 6.26 m s^{-1}
Bounce speed = $\sqrt{2 \times 9.81 \times 1.20}$ = 4.85 m s^{-1}
Taking upwards as positive:
Impact velocity = −6.26 m s^{-1}; bounce velocity = 4.85 m s^{-1}
Change of velocity, Δv = 4.85 −(−6.25) = 11.1 m s^{-1} upwards
∴ Change of momentum, $\Delta p = m\Delta v$ = 0.220 ×11.1= 2.44 N s = 2.4 N s (2 sf)

(b) Resultant force = $\dfrac{\Delta p}{t} = \dfrac{2.44 \text{ N s}}{0.150 \text{ s}}$ = 16.3 N

(c) There are two vertical forces on the ball: the weight, W (= 0.220 × 9.81 = 2.2 N), which acts downwards, and the contact force, C, by the ground, which acts upwards.
∴ Resultant force $R = C − W$, so $C = R + W = R + 2.2$ N.

Q10 (a) Take motion to the right as positive. Applying the principle of conservation of momentum, sum of mv terms is constant:
$$0.15 \times 0.36 - 0.25 \times 0.52 = 0.15v + 0.25 \times 0.107$$
where v = velocity of 0.15 kg glider after the collision.
∴ 0.15 v = −0.10275
∴ v = −0.685 m s^{-1}, i.e. 0.685 m s^{-1} (3 sf) to the left.

(b) Initial kinetic energy = $\frac{1}{2}$ × 0.15 × (0.36)^2 + $\frac{1}{2}$ × 0.25 × (0.52)^2 = 0.0435 J
Final kinetic energy = $\frac{1}{2}$ × 0.15 × (−0.685)^2 + $\frac{1}{2}$ × 0.25 × (0.107)^2 = 0.0366 J
∴ There is a loss of kinetic energy, hence inelastic.

Alternatively: Closing speed (before collision) = 0.36 + 0.52 = 0.88 m s^{-1}
Separating speed (after collision) = 0.685 + 0.107 = 0.792 m s^{-1}
0.792 m s^{-1} < 0.88 m s^{-1}, so inelastic.

Q11 (a) Normal component of velocity changes from −2500 cos 60° to +2500 cos 60°.
So Δv = 2500 m s^{-1}.
So $\Delta p = m\,\Delta v$ = 6.6 × 10^{-27} kg × 2500 m s^{-1} = 1.65 × 10^{-23} N s (upwards)

(b) To get back to side XY, molecule must travel vertically by 6.0 cm + 6.0 cm = 12.0 cm. The vertical component of velocity = ±1250 m s^{-1}.
∴ Time between collisions on XY = $\dfrac{0.120 \text{ m}}{1250 \text{ m s}^{-1}}$ = 9.6 × 10^{-5} s

∴ Mean force on XY = $\dfrac{1.65 \times 10^{-23} \text{ N s}}{9.6 \times 10^{-5} \text{ s}}$ = 1.7 × 10^{-19} N

Section 1.4: Energy concepts

Q1   (a) Power = energy per unit time = $\dfrac{\text{energy}}{\text{time}}$; so [energy] = [power] × [time]
The hour (h) is a unit of time, so [energy] = kW h

(b) 96 kW h = 96 × 10^3 W × 3 600 s
= 3.5 × 10^8 J (2 sf)

Q2 (a) Energy can neither be created nor destroyed but can be transferred from one form (or body) to another.

(b) (i) Initial kinetic energy $= \frac{1}{2}mv^2 = 0.5 \times 0.150 \times (50.0)^2 = 187.5$ J

Gain of gravitational potential energy $= mgh = 0.150 \times 9.81 \times 31.9 = 46.9$ J

$\therefore$ Kinetic energy at 31.9 m $= 187.5 - 46.9 = 140.6$ J

$\therefore 0.5 \times 0.150 \times v^2 = 140.6$

$\therefore$ Speed, $v = 43$ m s^{-1} (2 sf)

(ii) No, all the values of energy contain the factor $m = 0.150$ kg, so it could be changed or cancelled out without affecting the calculation.

Q3 (a) Loss in GPE $= 85$ kg $\times 9.81$ N kg$^{-1} \times 200$ m sin $5°$

$= 14\,500$ J

(b) Gain in kinetic energy $= \frac{1}{2}mv^2 = 0.5 \times 85$ kg $\times (12.0$ m s$^{-1})^2 = 6\,120$ J

$\therefore$ Loss of mechanical energy $= 14\,500$ J $- 6\,100$ J $= 8\,400$ J

$\therefore$ Work done against friction $= 8\,400$ J, so frictional force $\times 200$ m $= 8\,400$ J,

$\therefore$ Frictional force $= \dfrac{8\,400\,\text{J}}{200\,\text{m}} = 42$ N (2 sf)

Q4 (a) Loss of kinetic energy = work done against friction

$\therefore F \times 55$ m $= 0.5 \times 1\,200$ kg $\times (26.7$ m s$^{-1})^2$

$= 428\,000$ J

$\therefore F = 7\,800$ N (2 sf)

(b) $F_{30} = \dfrac{0.5 \times 1200 \times (13.35)^2}{14\,\text{m}} = 7\,600$ N (2 sf); $F_{50} = \dfrac{0.5 \times 1200 \times (22.25)^2}{38\,\text{m}} = 7\,800$ N (2 sf)

$F_{70} = \dfrac{0.5 \times 1200 \times (31.15)^2}{75\,\text{m}} = 7\,800$ N (2 sf) (Velocities converted with factor 60 mph $= 26.7$ m s^{-1})

All the forces are very close, so the assumption is true.

[Note: An alternative, and more insightful, method is to compare the ratios v^2/d and show that they are all very close. There is no need to convert mph to m s^{-1}.]

(c) John is incorrect. The kinetic energy loss is proportional to the mass of the car, so the braking force must also be proportional to the mass.

Q5 (a) The energy possessed by a system is defined as the amount of work it can do. When a system does 10 J of work, then 10 J of energy is transferred. Hence the two definitions are the same.

(b) In the case of the horse, the energy transfer is effected by a force moving its point of application. Hence the work definition is useful. In the case of the Sun, the energy transfer is from individual emissions of photons, a situation in which force is not a useful concept.

Q6 (a) Initial GPE $= mgh$

$= 0.600$ kg $\times 9.81$ N kg$^{-1} \times 0.400$ m $= 2.35$ J.

(b) Loss in GPE $= 0.600 \times 9.81 \times 0.184 = 1.08$ J

Elastic potential energy, EPE $= \frac{1}{2}kx^2 = 0.5 \times 32.0$ N m$^{-1} \times (0.184$ m$)^2 = 0.54$ J

Kinetic energy $= 1.08$ J $- 0.54$ J $= 0.54$ J

(c) $\frac{1}{2} \times 0.600$ kg $\times v^2 = 0.54$ J

$\therefore v = \sqrt{\dfrac{2 \times 0.54\,\text{J}}{0.600\,\text{kg}}} = 1.34$ m s^{-1} (3 sf)

(d) At the lowest point, KE $= 0$, so: loss of GPE = gain of EPE

If the distance fallen $= x$, $mgx = \frac{1}{2}kx^2$, $\therefore x = \dfrac{2mg}{k} = \dfrac{2 \times 0.600\,\text{kg} \times 9.81\,\text{N kg}^{-1}}{32\,\text{N m}^{-1}}$

$\therefore x = 0.368$ m

$\therefore$ Height above bench $= 0.400 - 0.368 = 0.032$ m

(e) (i) mgh = loss of GPE; $\frac{1}{2}kh^2$ = gain in EPE; $\frac{1}{2}mv^2$ = gain in KE (because initial KE $= 0$)

The equation is an expression of energy conservation: the loss of GPE is equal to the gain in EPE + the gain in KE

    (ii) Inserting values: $0.6 \times 9.81h = 16h^2 + 0.3$

        $\therefore 16h^2 - 5.886h + 0.3 = 0$

        $\therefore h = \dfrac{5.886 \pm \sqrt{5.886^2 - 4 \times 16 \times 0.3}}{2 \times 16} = 0.061$ m or 0.306 m

Q7 (a) From the definition of velocity, $x = vt$, where t = time.

    From the definition of power, $W = Pt$.

    Substituting for W and x in $W = Fx \longrightarrow Pt = Fvt$ and hence $P = Fv$.

  (b) (i) $[k] =$ N $($m s$^{-1})^{-2} =$ kg m s^{-2} (m^{-2} s^2)

            $=$ kg m^{-1}

    (ii) Velocity is constant, so drag D = forward force F

        $P = Fv = (kv^2)\,v$

               $= 0.4$ kg m$^{-1} \times (30$ m s$^{-1})^3$

               $= 10\ 800$ W

    (iii) [Note: in this question, your answer will depend very much on your assumptions]

        Assume a steady speed of 30 m s^{-1} and an efficiency of 80%

        Time taken for 900 km $= \dfrac{900 \times 10^3}{30}$ s $= 30\ 000$ s

        $\therefore$ Energy transfer $= 10\ 800$ W $\times 30\ 000$ s $= 3.2 \times 10^8$ J

        80% of 100 kW h $= 0.8 \times 100 \times 3.6 \times 10^6$ J $= 2.9 \times 10^8$ J.

        This suggests that the claim is incorrect but not very far off – perhaps the assumed driving speed was less.

Q8 (a) Ignoring the angle to the horizontal, $W = 83$ N $\times 7.0 \times 10^3$ m

                                         $= 5.8 \times 10^5$ J (2 sf)

  (b) Strictly, $W = Fx \cos \theta$, where θ = angle to the horizontal.

    But, if $\theta < 5°$, $\cos \theta > 0.9962$, so for an answer to 2 sf (like the data) the actual angle is irrelevant.

  (c) 83 N

  (d) The sledge is moving at constant velocity, so the resultant force on the sledge is zero, so there is no energy transfer to the sledge. Energy is transferred to the ice as thermal energy.

  (e) (i) Kinetic energy $\longrightarrow$ thermal energy (internal energy)

    (ii) Kinetic energy of sledge $= \frac{1}{2} \times 210$ kg $\times (1.4$ m s$^{-1})^2 = 205.8$ J

        Distance travelled $= \dfrac{\text{work done}}{\text{frictional force}} = \dfrac{\text{drop in kinetic energy}}{\text{frictional force}} = \dfrac{205.8 \text{ J}}{83 \text{ N}} = 2.5$ m (2 sf)

  (f) The resultant force $= 105$ N $- 83$ N $= 22$ N, so nearly 4 times as much of the work is done in overcoming friction as in accelerating the sledge.

Q9 (a) (i) [LHS] = kg s^{-1}

        [RHS] = m^2 (m s^{-1}) kg m$^{-3} =$ kg s$^{-1} =$ [LHS] , so homogeneous

    (ii) The power input is the KE input per second $= \frac{1}{2}mv^2$

                                        $= \frac{1}{2}\pi r^2 v \rho \times v^2$

                                        $= \frac{1}{2}\pi r^2 \rho v^3$

    (iii) $P_{\text{OUT}} = 0.56 \times 0.95 \times 0.85 \times \frac{1}{2}\pi \times (6.0$ m$)^2 \times 1.3$ kg m$^{-3} \times (15$ m s$^{-1})^3$

           $= 110\ 000$ W (2 sf)

           $= 110$ kW

  (b) The mean power output $\propto v^3$ so at half the wind speed, the power output would be one eighth of the value in (a)(iii), so this would not be enough. But to calculate the mean power output we need to know the mean cube speed, e.g. if on two days, the speeds were 0 and 15 m s^{-1}, the mean power output (for the two days) would be 55 kW, which would still be too small and Bethan is probably right.

Section 1.5: Circular motion

Q1 The angle θ, expressed in radians, is defined by: $\theta = \frac{l}{r}$ (see diagram). In this case $\frac{l}{r} = 1.2$.

The circumference of a circle $= 2\pi r$, so $360° = 2\pi$ rad,

i.e. $(\theta \: / \: °) = \frac{360}{2\pi} (\theta \: / \: \text{rad})$

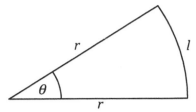

Q2 The period, T, is the time taken for one complete rotation. The frequency, f, is the number of rotations per unit time [or per second]. From these definitions:

If there are N rotations in a time t: $T = \frac{t}{N}$ and $f = \frac{N}{t}$. Hence $f = \frac{1}{T}$.

Q3 (a) [For an object travelling in a circle] the angular velocity is the angle swept out by the radius per unit time.

(b) $\omega = \frac{1400 \times 2\pi \text{ rad}}{60 \text{ s}} = 150$ rad s^{-1} (2 sf)

Q4 (a) Tension in rope = centripetal force on the object

$$= \frac{mv^2}{r} = \frac{65 \text{ kg} \times (23.2 \text{ m s}^{-1})^2}{4.5 \text{ m}}$$

$$= 7\,800 \text{ N (2 sf)}$$

(b) The rope does work on the object because the direction of motion (towards the centre) is the same as the direction of the force. Hence energy is transferred to the object. Hence its kinetic energy increases, i.e. it speeds up.

Q5 (a) Acceleration and resultant force are both directed towards the centre of the circle. The force is provided by the [sideways] grip of the road on the car tyres.

(b) $\frac{m(v_{max})^2}{r} = mg$, so $v_{max} = \sqrt{rg} = \sqrt{24.0 \times 9.81}$ m s^{-1} = 15.3 m s^{-1} (3 sf)

(c) From part (b) $v_{max} = \sqrt{rg}$, so as r increases so does v_{max} and this claim is correct.

$mr(\omega_{max})^2 = mg$, so $\omega_{max} = \sqrt{\frac{g}{r}}$, so as r increases ω_{max} decreases and this claim is also correct.

Q6 (a) (i) $\omega = \frac{2\pi \text{ rad}}{(29.5 \times 86\,400 \times 365.25) \text{ s}} = 6.75 \times 10^{-9}$ rad s^{-1}

(ii) $\omega = \frac{2\pi \text{ rad}}{(10.7 \times 3600) \text{ s}} = 1.63 \times 10^{-4}$ rad s^{-1}

(b) (i) Orbital speed, $v = r\omega = 1.43 \times 10^9$ km $\times 6.75 \times 10^{-9}$ rad s^{-1} = 9.65 km s^{-1}

(ii) Rotational speed, $v = r\omega = 60\,000$ km $\times 1.63 \times 10^{-4}$ rad s^{-1} = 9.78 km s^{-1}

(c) (i) Centripetal acceleration at equator $= r\omega^2 = 6.51 \times 10^{-5}$ m s^{-2} [or 6.52 – rounding]

Centripetal force $= ma = 3.70 \times 10^{22}$ N provided by the gravitational force of the Sun

(ii) Centripetal acceleration $= r\omega^2$

$$= 1.59 \text{ m s}^{-2}$$

∴ Percentage reduction in measured $g = \frac{1.59 \text{ m s}^{-2}}{10.4 \text{ m s}^{-2}} \times 100\% = 15\%$ (2 sf)

Q7 (a) The centripetal force is provided by the horizontal component of the tension in the pendulum string.

(b) Vertical component of tension, T = weight of the bob

∴ $T \cos 20° = 0.078$ kg $\times 9.81$ N kg^{-1}

∴ $T = 0.8143$ N = 0.81 N (2 sf)

$T \sin 20° = \frac{mv^2}{r}$, so $v = \sqrt{\frac{0.153 \text{ m} \times 0.8143 \text{ N} \times \sin 20°}{0.078 \text{ kg}}} = 0.739$ m s^{-1}

Alternative answer:
Resolve vertically: $T \cos \theta = mg$, so $T = \frac{mg}{\cos \theta}$

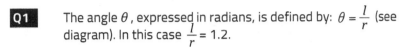

Component 1 Answers

Horizontally: $T\sin\theta = \dfrac{mv^2}{r}$. Substituting for $T \longrightarrow \dfrac{mg\sin\theta}{\cos\theta} = \dfrac{mv^2}{r}$

∴ Simplifying: $v = \sqrt{rg\tan\theta} = \sqrt{0.153\text{ cm}\times 9.81\text{ m s}^{-2}\times\tan 20°} = 0.739\text{ m s}^{-1}$

Q8 (a) The (inward) gravitational force on the smaller body $= \dfrac{GMm}{r^2}$, where m is the mass of the less massive object. This provides the centripetal force, $\dfrac{mv^2}{r}$

So $\dfrac{mv^2}{r} = \dfrac{GMm}{r^2}$. Multiplying by r and dividing by $m \longrightarrow v^2 = \dfrac{GM}{r}$ QED

(b) Without dark matter $v^2 \propto \dfrac{1}{r}$, so $\left(\dfrac{v_1}{v_2}\right)^2 = \dfrac{r_2}{r_1} =$ constant.

$\left(\dfrac{4700}{3400}\right)^2 = 1.9$. This is very close to 2, so this result is consistent with the absence of dark matter.

This is just a single galaxy, so the data are interesting but not definitive.

Section 1.6: Vibrations

Q1 (a) (i) 0.040 m

(ii) $\omega = \dfrac{2\pi}{T} = \dfrac{2\pi}{1.20\text{ s}} = 5.24\text{ rad s}^{-1}$

(iii) $-\dfrac{\pi}{2}$

(b) $v_{max} = A\omega = 0.040\text{ m}\times 5.24\text{ rad s}^{-1} = 0.21\text{ m s}^{-1}$
Initial velocity is maximum positive, so a cosine graph.

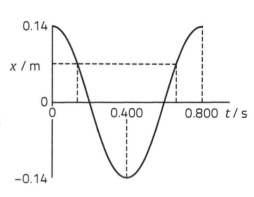

Q2 (a) (i) $x = A\cos\omega t = 0.140\times\cos\left(\dfrac{2\pi}{0.800\text{ s}}\times 0.50\text{ s}\right)$

$= -0.099\text{ m}$

(ii) $0.070\text{ m} = \frac{1}{2}A$, so $\cos\omega t = 0.5$ and $\omega t = \dfrac{\pi}{3}$

So $\dfrac{2\pi}{0.800\text{ s}}t = \dfrac{\pi}{3}$, ∴ $t = 0.133\text{ s}$

From the graph, the 2nd occasion $= 0.800 - 0.133\text{ s}$
$= 0.667\text{ s}$

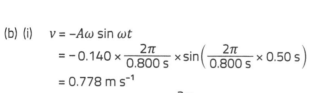

(b) (i) $v = -A\omega\sin\omega t$

$= -0.140\times\dfrac{2\pi}{0.800\text{ s}}\times\sin\left(\dfrac{2\pi}{0.800\text{ s}}\times 0.50\text{ s}\right)$

$= 0.778\text{ m s}^{-1}$

(ii) $v_{max} = A\omega = 0.140\times\dfrac{2\pi}{0.800\text{ s}} = 1.10\text{ m s}^{-1}$.

∴ $\sin\omega t = 0.5$, ∴ $\dfrac{2\pi}{0.800}t = \dfrac{\pi}{6}$, ∴ $t = 0.067\text{ s}$

From the graph, the 2nd occasion $= 0.400 - 0.067\text{ s}$
$= 0.333\text{ s}$

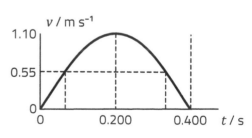

Q3 (a) The (stiffness of the) ruler.

(b) $\omega = 2\pi f = \sqrt{\dfrac{k}{m}}$, $\therefore k = 4\pi^2 f^2 m = 4\pi^2 \times (0.40 \text{ Hz})^2 \times 0.20 \text{ kg} = 1.3 \text{ N m}^{-1}$ [or kg s^{-2}]

(c) $v_{\max} = A\omega = 2\pi A f$

$= 2\pi \times 0.050 \text{ m} \times 0.40 \text{ Hz}$

$= 0.13 \text{ m s}^{-1}$

$T = \dfrac{1}{0.4 \text{ Hz}} = 2.5 \text{ s}$

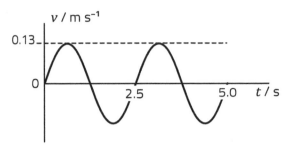

Q4 **Algebraic method**: There are various ways of setting out. This is a compact one:

$T = 2\pi\sqrt{\dfrac{l}{g}}$, so $g_{\text{moon}} = g_{\text{Earth}} \times \left(\dfrac{T_{\text{Earth}}}{T_{\text{moon}}}\right)^2 = 9.81 \times \left(\dfrac{100}{240}\right)^2 = 1.70 \text{ m s}^{-2}$ (3 sf)

Alternative method: Calculate length of pendulum on Earth $\longrightarrow 0.248$ m. Then use this to calculate g on Moon from the period $\longrightarrow \surd 1.70 \text{ m s}^{-2}$.

Q5 (a) Time taken = ½ × period = $0.5 \times 2\pi\sqrt{\dfrac{m}{k}} = 0.5 \times 2\pi\sqrt{\dfrac{0.200 \text{ kg}}{40 \text{ N m}^{-1}}} = 0.22 \text{ s}$ (2 sf)

(b) It is true that there is a larger resultant upward force at the maximum extension but there is a larger distance to travel. The period of a body undergoing SHM is independent of the amplitude, e.g. for a mass on a spring it depends only on the mass and the stiffness: $T = 2\pi\sqrt{m/k}$, and the time involved is half the period, so Fergus's conclusion is invalid.

Q6 The spring constant $k = \dfrac{mg}{l}$, where m is the mass of object, so $\dfrac{m}{k} = \dfrac{l}{g}$.

The period, T, of oscillation of the object on the spring is given by $T = 2\pi\sqrt{\dfrac{m}{k}}$.

$\therefore$ Substituting for $\dfrac{m}{k} \longrightarrow T = 2\pi\sqrt{\dfrac{l}{g}}$. But this is just the same as the period of a simple pendulum of length l, so it is not a fluke and Davinder is (on this occasion) incorrect.

Q7 (a) If a pendulum of length l is displaced by angle θ, the distance d of the bob below the pivot is given by:

$d = l \cos \theta$.

So the height raised, $h = l(1 - \cos \theta)$

$\therefore$ Gain in GPE $= mgl(1 - \cos \theta)$

$= 0.100 \times 9.81 \times 1.00 (1 - \cos 11.5°)$

$= 0.0197 \text{ J} \sim 0.02 \text{ J}$

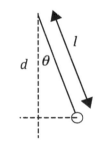

(b) (i)

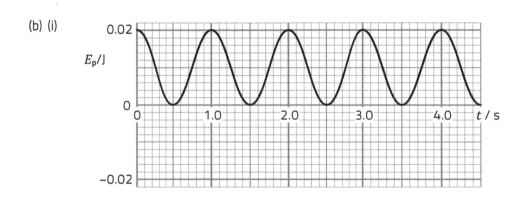

(ii)

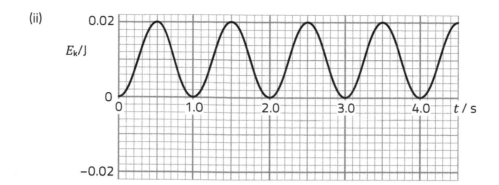

Q8 (a) Air resistance will act on the mass and disc. This is a resistive force and always opposes the motion. Energy will be lost to random kinetic energy of air molecules, i.e. internal energy of the air will increase slightly.

(b) From 0.15 s to 1.95 s the amplitude halves (from 0.2 m s^{-1} to 0.1 m s^{-1}), i.e. a half-life of 1.80 s. Also halves (0.1 to 0.05) from 1.95 s to 3.75 s. Again a half-life of 1.80 s.
Final check: another half-life is 0.75 s to 2.55 s (0.16 to 0.08). Another half-life of 1.80 s.
Constant half-life means an exponential decay and Sophie is therefore correct.

(c) Velocity halves so the KE will reduce to one quarter, since KE = $\frac{1}{2}mv^2$
Hence 75% of the KE has been lost or dissipated.

Q9 (a) Critical damping is when the resistive forces are just large enough to stop oscillations occurring.

(b) Car suspension. If oscillations took place, the tyres could leave the road, resulting in less friction and increased braking distances. This would increase the number of accidents occurring (this can have an enormous effect on a bumpy road).

Q10 (a) Forced oscillations is when a periodic driving force acts on an oscillatory system. The system then oscillates at the frequency of the driving force.

(b) (i) It is the periodic (or sinusoidal) force applied to it as a 'push and pull' by the vibrating pin via the weak spring.

(ii) From the graph, the resonant frequency is 0.8 Hz and this must be the natural frequency of the pendulum. Even though there is a fair amount of damping, the frequency will only decrease a very tiny percentage away from the natural frequency.

Period, $T = \dfrac{1}{0.80 \text{ Hz}} = 1.25$ s.

Rearranging $T = 2\pi\sqrt{\dfrac{l}{g}} \longrightarrow l = \left(\dfrac{T}{2\pi}\right)^2 g = \left(\dfrac{1.25}{2\pi}\right)^2 \times 9.81 = 0.39$ m (2 sf)

Section 1.7: Kinetic theory

Q1 Gases consist of a large number of molecules in random motion in (otherwise) empty space.
The volume of the molecules themselves is a negligible fraction of the volume of the gas.
Collisions between molecules are perfectly elastic and take a negligible time.
The molecules exert negligible forces on one another except during collision.

Q2 The mole is the amount of a substance with as many particles as there are in exactly 12 g of carbon-12.
The Avogadro constant is the number of particles per mole [~6.02 × 10^{23} mol^{-1}]

[Note that these are the definitions from the Terms and Definitions booklet. They are out of date. Since 2019 the mole has been defined as exactly 6.022 140 76 × 10^{23} particles and hence the Avogadro constant, N_A, is 6.022 140 76 × 10^{23} mol^{-1}. Either pair of definitions will be credited in Eduqas examinations.]

Q3 The molecules of a gas, in their random rapid motion, collide with the walls of the container. In such a collision, a molecule suffers a change in momentum inwards at right angles (on the average) to the wall. There are a large number of such collisions per second on any area of wall, so the wall exerts a force on the molecules (equal to the change in momentum per second − Newton's second law). By Newton's third law, the molecules exert an equal and opposite force on the wall. The pressure is the magnitude of this force divided by the area of the wall.

If the temperature is increased, the molecular speeds increase, so there are more collisions per second with the wall and each of them results in a larger change of momentum. Hence, the pressure exerted by the gas molecules increases with temperature.

Q4 The ideal gas equation of state is $pV = nRT$, where n is the amount of the gas, ie. the number of moles. The equation can also be expressed as $pV = NkT$, where N is the number of molecules of the gas. $N = nN_A$, so the second equation can be written $pV = nN_A kT$.

Comparing this with the first equation, $nN_A k = nR$. Hence $k = \dfrac{R}{N_A}$.

Q5 For a single molecule, KE $= \frac{1}{2}mc^2$, so the mean molecular kinetic energy $= \frac{1}{2}m\overline{c^2}$ and the total kinetic energy in a gas, $U = \frac{1}{2}Nm\overline{c^2}$, where N is the number of molecules.
Equating the right-hand sides of the given equations: $\frac{1}{3}Nm\overline{c^2} = nRT$
But $\frac{1}{3}Nm\overline{c^2} = \frac{2}{3}U$, $\therefore \frac{2}{3}U = nRT$, and hence $U = \frac{3}{2}nRT$

[Note: this expression is only valid for the translational kinetic energy of molecules − it doesn't include any contribution from rotational kinetic energy which is significant in gases with polyatomic molecules.]

Q6 $pV = \frac{1}{3}Nm\overline{c^2}$

so $V = \dfrac{Nm\overline{c^2}}{3p} = \dfrac{\text{mass of gas} \times \overline{c^2}}{3p}$

$= \dfrac{3.0\,\text{mol} \times 0.028\,\text{kg mol}^{-1} \times (550\,\text{m s}^{-1})^2}{3 \times 140 \times 10^3\,\text{Pa}}$

$= 0.061\,\text{m}^3$

Q7 (a) $pV = nRT$

so $n = \dfrac{pV}{RT} = \dfrac{102 \times 10^3\,\text{Pa} \times 0.89\,\text{m}^3}{8.31\,\text{J mol}^{-1}\,\text{K}^{-1} \times 298\,\text{K}} = 36.7\,\text{mol}$, that is 37 mol (to 2 sf)

(b) $\frac{1}{2}m\overline{c^2} = \frac{3}{2}kT$

so $c_\text{rms} = \sqrt{\overline{c^2}} = \sqrt{\dfrac{3kT}{m}} = \sqrt{\dfrac{3 \times 1.38 \times 10^{-23}\,\text{J K}^{-1} \times 298\,\text{K}}{6.64 \times 10^{-27}\,\text{kg}}} = 1\,360\,\text{m s}^{-1}$

(c) $pV = nRT$

so $V = \dfrac{nRT}{p} = \dfrac{36.7\,\text{mol} \times 8.31\,\text{J mol}^{-1}\,\text{K}^{-1} \times 232\,\text{K}}{23 \times 10^3\,\text{Pa}} = 3.1\,\text{m}^3$

We are assuming that (i) the inward pressure exerted by the stretched balloon skin itself is negligible compared with the 23 kPa atmospheric pressure, (ii) all the helium has reached 232 K, (iii) the helium is behaving as an ideal gas. [Note: the examiner would expect one assumption and anticipate the first of these.]

Q8 (a) The total number of moles stays constant. Calculating it from the initial data:

$n = \left(\dfrac{pV}{RT}\right)_\text{left} + \left(\dfrac{pV}{RT}\right)_\text{right}$

$= \dfrac{1.02 \times 10^5\,\text{Pa} \times 37.0 \times 10^{-3}\,\text{m}^3}{8.31\,\text{J mol}^{-1} \times 293\,\text{K}} + \dfrac{6.50 \times 10^5\,\text{Pa} \times 22.5 \times 10^{-3}\,\text{m}^3}{8.31\,\text{J mol}^{-1} \times 293\,\text{K}}$

$= 1.55\,\text{mol} + 6.00\,\text{mol} = 7.55\,\text{mol}$

After thermal equilibrium, this number of moles occupies $59.5 \times 10^{-3}\,\text{m}^3$ at 293 K,

so $p = \dfrac{nRT}{V} = \dfrac{7.55\,\text{mol} \times 8.31\,\text{J mol}^{-1}\,\text{K}^{-1} \times 293\,\text{K}}{59.5 \times 10^{-3}\,\text{m}^3}$

$= 3.1 \times 10^5\,\text{Pa}$

(b) [Assuming opening of tap is equivalent to release of light gas-tight piston separating the two containers and that the process is quick enough for heat flow to be negligible.] High pressure gas on the right will do work on the gas on the left by compressing it, lose internal energy and cool; low pressure gas on the left will have work done on it and become warmer. So Tudor is correct.

Q9 (a) $p = \frac{1}{3}\rho\overline{c^2}$

so $c_{rms} = \sqrt{\overline{c^2}} = \sqrt{\frac{3p}{\rho}} = \sqrt{\frac{3 \times 112 \times 10^3 \text{ Pa}}{1.35 \text{ kg m}^{-3}}} = 499 \text{ m s}^{-1}$

(b) Mass of gas = density × volume = ρV

But $n = \dfrac{pV}{RT}$

So mass per mole = $\dfrac{pV}{n} = \dfrac{\rho RT}{p} = \dfrac{1.35 \text{ kg m}^{-3} \times 8.31 \text{ J mol}^{-1} \text{ K}^{-1} \times 293 \text{ K}}{112 \times 10^3 \text{ Pa}}$

$= 0.0293 \text{ kg mol}^{-1}$

(c) (i) $n = \dfrac{pV}{RT} = \dfrac{935 \times 10^3 \text{ Pa} \times 1.5 \times 10^{-3} \text{ m}^3}{8.31 \text{ J mol}^{-1} \text{ K}^{-1} \times 320 \text{ K}} = 0.527 \text{ mol}$

So mass of air in bottle = $0.527 \times 10^{-3} \text{ mol} \times 0.0293 \text{ kg mol}^{-1}$
$= 0.015 \text{ kg}$

Assumption: the air is an ideal gas.

(ii) Work is being done on the gas as it is pumped in. Not much heat escapes, so the internal energy of the gas rises, hence its temperature increases.

Section 1.8: Thermal physics

Q1 The internal energy of a system is the sum of the kinetic and potential energies of the particles in the system.

Q2 At absolute zero, the internal energy of a system is the minimum possible. It is impossible to extract energy from the sy stem.

Q3 (a) Unlike in other systems, the molecules of an ideal gas do not exert forces on each other, so there is no potential energy component to the internal energy – it is entirely kinetic energy.

(b) $U = \frac{3}{2}NkT$. $N = N_A \times \dfrac{30 \text{ g}}{20 \text{ g}} = 1.5N_A$, $T = (273.15 + 26.85) \text{ K} = 300 \text{ K}$

∴ $U = \frac{3}{2} \times 1.5 \times 6.022 \times 10^{23} \times 1.38 \times 10^{-23} \text{ J K}^{-1} \times 300 \text{ K} = 5600 \text{ J}$ (2 sf)

Q4 Heat is the flow of energy from one system to another due to a temperature difference. The flow is into the system with a lower temperature.

Q5 Thermal equilibrium means that there is no heat flow between the systems – their temperatures are the same.

Q6 (a) ΔU = the increase in internal energy of the system
Q = the heat flow into the system
W = the work done by the system
Q leads to a gain in internal energy of the system and W a loss. The net gain in internal energy (ΔU) is equal to the energy input due to heat minus the energy loss due to work.

(b) $W = p\Delta V$ where p is the pressure and ΔV the change in volume of the system. The volume of a solid or a liquid is almost constant, so $\Delta V \sim 0$. Hence $W = 0$.

Q7 The specific heat capacity of a substance is the heat required to raise the temperature of the substance, per unit mass per unit temperature rise.

Q8 The apparatus is set up without the Bunsen burner and with cold water and ice (at 0°C) in the beaker. The temperature is measured using the thermometer and the pressure of the air in the flask measured using the pressure gauge. The temperature of the water is raised using the Bunsen in a series of steps of approximately 10°C up to 100°C. At each step the Bunsen is removed, the water stirred, and time allowed for the system to equilibrate (stirring will help); the pressure of the air column and the temperature are measured.

The pressure of the air is plotted against the temperature (in°C) and a best fit straight line drawn. The intercept of this line on the temperature axis is the estimate of absolute zero.

Q9 First law of thermodynamics: $\Delta U = Q - W$.
$Q = 0$ so $\Delta U = -W$. The gas expands, so $W > 0$. Hence $\Delta U < 0$.
$U = \frac{3}{2}NkT$. N and k are constant so, if $\Delta U < 0$, $\Delta T < 0$.

Q10 (a) $W = p\Delta V = 1.42 \times 10^5\,\text{Pa} \times 2.7 \times 10^{-3}\,\text{m}^3$
$\quad = 380\,\text{J (2 sf)}$

(b) $W = p\Delta V = -380\,\text{J} + 1.42 \times 10^5\,\text{Pa} \times (-1.5 \times 10^{-3}\,\text{m}^3)$
$\quad = -590\,\text{J (2 sf)}$

Q11 Assumptions: Heat exchange with the environment is negligible; the heat capacity of the container is negligible. In this case:

Loss of internal energy of water = Gain of internal energy of carrots

Let θ = equilibrium temperature:

$\therefore 1.2\,\text{kg} \times 4210\,\text{J kg}^{-1}\text{°C}^{-1} \times (100°\text{C} - \theta) = 0.700\,\text{kg} \times 1880\,\text{J kg}^{-1}\text{°C}^{-1} \times (\theta - 20°\text{C})$
[Omitting the units for clarity]
$\therefore 505\,200 - 5052\theta = 1316\theta - 26\,320$
$\therefore 6368\theta = 531\,520$
$\therefore \theta = \dfrac{531\,520}{6368}°\text{C} = 83°\text{C (2 sf)}$

Q12 (a) (i) 0 [because the volume is constant]

(ii) $W = p\Delta V = 0.85 \times 10^5\,\text{Pa} \times (7.1 - 16.7) \times 10^{-3}\,\text{m}$
$\quad\quad = -816\,\text{J}$

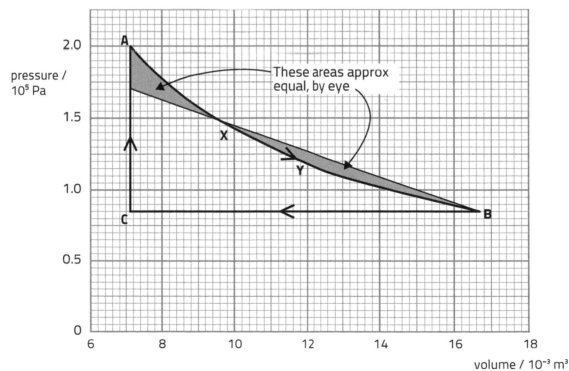

(iii) W = Area under graph A⟶C = Area under estimated straight line (see diagram)

= area of a trapezium

$= \frac{1}{2} \times (1.70 + 0.85) \times 10^5 \times (16.7 - 7.1) \times 10^{-3}$

= 1224 J

(b) If T is constant, pV is a constant:

[Omitting the factor $10^5 \times 10^{-3} = 100$ as this is only a comparison]

Checking values: $(pV)_A = 2.0 \times 7.1 = 14.2$; $(pV)_B = 0.85 \times 16.7 = 14.2$

$(pV)_X = 1.42 \times 10 = 14.2$; $(pV)_Y = 1.18 \times 12 = 14.2$

Hence the values of pV are the same at these four places suggesting that the temperature is constant along AB.

(c)

	AB	BC	CA	ABCA
ΔU / J	0	−1225	1225	0
Q / J	1200	−2045	1225	380
W / J	(a)(iii) 1200	(a)(ii) −820	(a)(i) 0	380

Some calculations: Using $U = \frac{3}{2}pV$, $U_A = U_B = \frac{3}{2} \times 14.2 \times 10^2$ J = 2130 J;

$U_C = \frac{3}{2} \times 0.85 \times 10^5$ Pa $\times 7.1 \times 10^{-3}$ m³ = 905 J

∴BC: $\Delta U = 905 - 2130 = -1225$ J, ∴ $Q = \Delta U + W = -1225 - 820 = -2045$ J

and CA: $(\Delta U)_{CA} = -(\Delta U)_{BC} = 1225$ J, ∴ $Q = \Delta U + W = 1225 + 0 = 1225$ J

Q13 Calculations: Heat supplied each minute = 12.00 V × 4.20 A × 60 s = 3024 J

∴ Temperature rise per minute $= \frac{Q}{mc} = \frac{3024 \text{ J}}{1.00 \text{ kg} \times 900 \text{ J kg}^{-1}{}^{\circ}\text{C}^{-1}} = 3.36\,^{\circ}\text{C}$

Suppose heat only starts reaching the thermometer by conduction after about 1 minute [Note: you wouldn't be penalised assuming the graph started rising straight away]

⟶ Potential temperature rise in 20 min = 3.36 × 19 = 63.8°C ⟶ top temp = 84°C

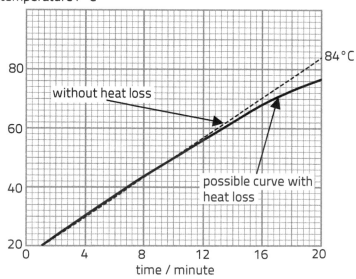

Q14 (a) $[p\,\Delta V] = [p][\Delta V] = (\text{N m}^{-2})\,\text{m}^3 = \text{N m} = \text{J} = [W]$ QED

(b) (i) **Method 1**: (calculating n)

Initially, $n = \frac{pV}{RT} = \frac{100 \times 10^3 \text{ Pa} \times 110 \times 10^{-6} \text{ m}^3}{8.31 \text{ J mol}^{-1} \text{K}^{-1} \times 293 \text{ K}} = 4.52$ mmol (to 3 sf)

Finally, $n = \frac{pV}{RT} = \frac{100 \times 10^3 \text{ Pa} \times 140 \times 10^{-6} \text{ m}^3}{8.31 \text{ J mol}^{-1} \text{K}^{-1} \times 373 \text{ K}} = 4.52$ mmol (to 3 sf)

The final amount of gas is the same as the initial to 3 sf, so a negligible amount has escaped.

Method 2: (not calculating n)
If the amount of gas and the pressure are constant, $V \propto T$:

$$\left(\frac{V}{T}\right)_{\text{initial}} = \frac{110 \times 10^{-6}\,\text{m}^3}{293\,\text{K}} = 3.75 \times 10^{-7}\,\text{m3 K}^{-1}\ \text{(to 3 sf)}$$

$$\left(\frac{V}{T}\right)_{\text{final}} = \frac{140 \times 10^{-6}\,\text{m}^3}{373\,\text{K}} = 3.75 \times 10^{-7}\,\text{m3 K}^{-1}\ \text{(to 3 sf)}$$

These are the same to 3 sf, so the amount escaping is negligible.

(ii) The work done by the gas, $W = p\Delta V = 100 \times 10^3\,\text{Pa} \times (140 - 110) \times 10^{-6}\,\text{m3} = 3.0\,\text{J}$

$\Delta U = \frac{3}{2} \times nR\Delta T = \frac{3}{2} \times 4.52 \times 10^{-3}\,\text{mol} \times 8.31\,\text{J mol}^{-1}\,\text{K}^{-1} \times (100 - 20)\,\text{K}$

$\qquad = 4.50\,\text{J}$

[Or: $\Delta U = \frac{3}{2} \times nR\Delta T = \Delta U = \frac{3}{2} \times p\Delta V$ (because p is constant) $= \frac{3}{2}W = 4.5\,\text{J}$]

From the first law of thermodynamics: $Q = \Delta U + W = 3.0\,\text{J} + 4.5\,\text{J} = 7.5\,\text{J}$

(iii) From the above numbers:
Heat needed to raise the temperature / mol and / degree $= \dfrac{7.5\,\text{J}}{4.52 \times 10^{-3}\,\text{mol} \times 80\,\text{K}}$

$$= 21\,\text{J mol}^{-1}\,\text{K}^{-1}$$

This is in agreement with Lucia's statement. However, if the gas were kept at a fixed volume, no work would be done by the gas, so the heat needed to raise the internal energy by the same amount would be less (because the internal energy of an ideal gas depends only on its temperature). So the heat needed isn't always 21 J.

Practice questions: **Component 2: Electricity and the Universe**

Section 2.1: **Conduction of electricity**

Q1 $Q = It = 0.015\,\text{A} \times 60\,\text{s} = 0.90\,\text{C}$

No. of electrons $= \dfrac{\text{charge}}{e} = \dfrac{0.90\,\text{C}}{1.60 \times 10^{-19}\,\text{C}} = 5.6 \times 10^{18}\ \text{(2 sf)}$

Q2 [An electrical conductor is] a material which allows charge to flow through it.

Q3 Re-arranging the equation: $C = \dfrac{Q^2}{2W}$; the number 2 has no unit.

$$\therefore F = [C] = \frac{[Q^2]}{[W]} = \frac{[Q]^2}{[W]} = \frac{\text{A}^2\,\text{s}^2}{\text{kg m}^2\,\text{s}^{-2}} = \text{kg}^{-1}\,\text{m}^{-2}\,\text{s}^4\,\text{A}^2$$

Q4 Charge on an α-particle $= 2e = 3.20 \times 10^{-19}\,\text{C}$

$\therefore$ Current $= 3.20 \times 10^{-19}\,\text{C} \times 37 \times 10^3\,\text{s}^{-1} = 1.2 \times 10^{-14}\,\text{A}$

Q5 $v = \dfrac{I}{nAe}\ ,\ \therefore\ \dfrac{v_A}{v_B} = \dfrac{I_A}{I_B} \times \dfrac{n_B}{n_A} \times \dfrac{A_B}{A_A} = \dfrac{I_A}{I_B} \times \dfrac{n_B}{n_A} \times \left(\dfrac{d_B}{d_A}\right)^2$

$\dfrac{v_A}{v_B} = \dfrac{1.5}{10} \times \dfrac{1.0 \times 10^{28}}{3.0 \times 10^{28}} \times \left(\dfrac{0.30}{0.60}\right)^2 = 0.013\ \text{(2 sf)}$

Q6 (a) The intensity of the light at the LDR is inversely proportional to the distance^2, so the number of photons reaching the LDR per second $\propto$ distance^{-2}. Assuming the number of conduction electrons is proportional to the number of photons per second, Nigel is correct.

(b)

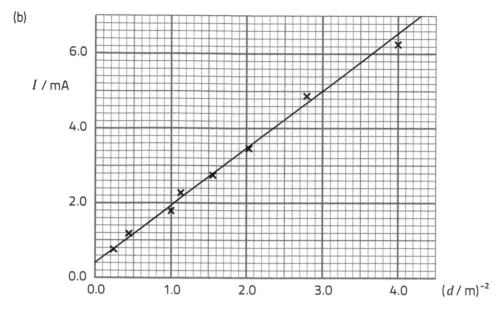

The graph of I against d^{-2} is expected to be a straight line through the origin. The points lie on quite a good straight line of positive gradient with a low degree of scatter, which agrees with Nigel's suggestion. However, the line does not pass through the origin. A possible reason for that is that the experiment was carried out in a room with some background light.

Section 2.2: Resistance

Q1 Current = rate of charge flow: $= I = \dfrac{Q}{t}, \therefore Q = It.$

pd = energy per unit charge flow $= \dfrac{W}{Q} = \dfrac{W}{It} = \dfrac{300\ \text{J}}{1.5\ \text{A} \times 20\ \text{s}} = 10\ \text{V}$

Q2 Current, $I = \dfrac{Q}{t}, \therefore [Q] = [I][t] = \text{A s}$

pd = energy per unit charge flow, so $V = \text{J C}^{-1} = (\text{kg m}^2\ \text{s}^{-2})\,(\text{A s})^{-1} = \text{kg m}^2\ \text{s}^{-3}\ \text{A}^{-1}$

Q3 (a) For a metallic conductor at a constant temperature the current is directly proportional to the potential difference.

(b) The equation $V = IR$ is a statement of Ohm's law only if it is also stated that the resistance, R, is a constant. Also the conditions should be stated (metallic conductor and constant temperature).

Q4 (a) n = number of free electrons per unit volume; A = cross-sectional area of conductor
e = electronic charge; v = free-electron drift velocity

(b) If the temperature of the wire increases, so does the mean kinetic energy of the free electrons. Their random speeds are therefore greater and so the time between collisions with the lattice ions is less. Because of this the increase in velocity due to the pd is less than at lower temperatures and so is the drift velocity, making the current less. A lower current for the same pd means that the resistance is greater.

Q5 (a) Superconductivity is the property of zero electrical resistance, which occurs when the conductor is below a certain temperature – the transition temperature , T_c.

(b) A high-temperature superconductor is one with a transition temperature above the boiling point of liquid nitrogen – about $-200\,°\text{C}$.

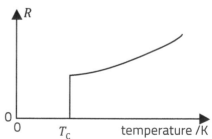

Advantage: it costs much less to achieve a superconducting state, using liquid nitrogen than conventional superconductors, which require liquid helium for cooling, especially in high-field industrial magnets such as in MRI.

Q6 (a) (i) $R_{2.4V} = \dfrac{2.4\ V}{1.5\ A} = 1.6\ \Omega;$ $\qquad R_{12V} = \dfrac{12\ V}{3.0\ A} = 4.0\ \Omega$

$\therefore \dfrac{\text{Resistance of the filament at 12 V}}{\text{Resistance of the filament at 2.4 V}} = 2.5$

(ii) $\dfrac{P_{12V}}{P_{2.4V}} = \dfrac{12\ V \times 3.0\ A}{2.4\ V \times 1.5\ A} = 10$

(b) As the pd increases, the free electrons have a greater acceleration between collisions, so gain more kinetic energy. Hence more energy is transferred to the lattice ions when they collide.

Q7 (a) Consider 1.00 m of wire.

$\rho = \dfrac{RA}{l} = \dfrac{2.18\ \Omega}{1.00\ m} \times \pi \left(\dfrac{0.1 \times 10^{-3}\ m}{2} \right)^2 = 1.7 \times 10^{-8}\ \Omega\ m$

(b) Volume of 1 kg (14 306 m length) of wire $= \pi \left(\dfrac{0.1 \times 10^{-3}\ m}{2} \right)^2 \times 14\ 306\ m = 1.12 \times 10^{-4}\ m^3$

$\therefore$ Density $= \dfrac{mass}{volume} = \dfrac{1.00\ kg}{1.12 \times 10^{-4}\ m^3} = 8900\ kg\ m^{-3} = 8.9\ g\ cm^{-3}$, so in good agreement.

Q8 (a) $\rho = \dfrac{RA}{l} = \dfrac{13.9\ \Omega}{2.000\ m} \times \pi \left(\dfrac{0.32 \times 10^{-3}\ m}{2} \right)^2 = 5.589 \times 10^{-7}\ \Omega\ m$

Fractional uncertainties: $p_d = \dfrac{0.01}{0.32} = 0.03125, \therefore p_A = 2 \times 0.03125 = 0.0625$

$p_l = \dfrac{0.002}{2.0} = 0.001;\ p_R = \dfrac{0.1}{13.9} = 0.007 \therefore p_\rho = 0.0625 + 0.001 + 0.007 = 0.0705$

$\therefore$ Absolute uncertainty: $\Delta_\rho = 0.0705 \times 5.589 \times 10^{-7} = 0.4 \times 10^{-7}$

$\therefore \rho = (5.6 \pm 0.4) \times 10^{-7}\ \Omega\ m$

(b) For double the diameter the p_d is half and p_A is one quarter of the new value, so ~0.031. The resistance would also be about one quarter, so with the same uncertainty in R, p_R would be 4 times the original, that is 0.028 and so the total p would be about 0.06, which is slightly, but only slightly, less. The uncertainty in length hardly affects the total uncertainty, but the fractional uncertainty in resistance will be halved (from 0.007 to 0.0035) which has a bigger effect.

Q9 $P = \dfrac{V^2}{R}$, so, when operating, $R = \dfrac{V^2}{P} = \dfrac{(240\ V)^2}{60\ W} = 960\ \Omega.$

Taking room temperature to be 290 K: operating temperature $= \dfrac{960\ \Omega}{80\ \Omega} \times 290\ K = 3500\ K$ (2 sf)

Q10 (a) Because the resistance is constant, the current taken from the supply and therefore the power of the heater will also be constant.

(b) $P = \dfrac{V^2}{R}$, so $R = \dfrac{V^2}{P} = \dfrac{(30\ V)^2}{10\ W} = 90\ \Omega$

$R = \dfrac{\rho l}{A}$, so $l = \dfrac{RA}{\rho} = \dfrac{90\ \Omega}{4.9 \times 10^{-7}\ \Omega\ m} \times \pi \left(\dfrac{0.12 \times 10^{-3}\ m}{2} \right)^2 = 2.1\ m$ (2 sf)

Q11 $R_A = \dfrac{4\rho l}{\pi D^2}$

$R_B = \dfrac{4(2.5\rho)(3l)}{\pi (2D)^2} = \dfrac{7.5}{4} \times \dfrac{4\rho l}{\pi D^2} = 1.875\ R_A$

$P = \dfrac{V^2}{R}$. The voltages are the same, so $\dfrac{P_A}{P_B} = \dfrac{R_B}{R_A} = 1.875$

Q12 (a) The wire coil is connected to a multimeter on its resistance range. The test tube is placed in a beaker with a water / ice mixture and allowed to equilibrate. A thermometer is placed in the beaker and the resistance and temperature is measured. The beaker is gradually heated using a bunsen burner and the resistance measured at a range of temperatures up to 100 °C.

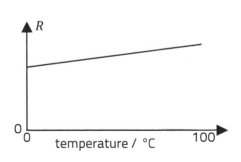

(b) Assuming that the resistance varies linearly with temperature:

$$\text{Increase in resistance per degree} = \frac{16.3\ \Omega - 12.0\ \Omega}{75\,°C - 20\,°C} = 0.0782\ \Omega\,°C^{-1}$$

$$\therefore \text{Temperature of oil} = 20\,°C + \frac{18.7\ \Omega - 12.0\ \Omega}{0.0782\ \Omega\,K^{-1}} = 85.7\ K$$

so temperature of oil = 20 °C + 85.7 °C = 106 °C

Q13

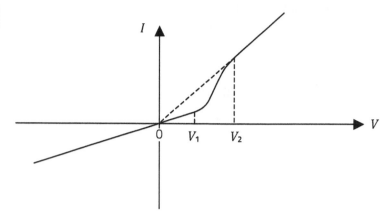

Section 2.3: DC circuits

Q1 (a) With the switch open, the lamps L_1 and L_2 are in series. Hence, the current is the same in the two lamps, so they have the same brightness.

(b) When S is closed, L_2 and L_3 are in parallel, so the resistance of this combination is less than that of L_2 on its own. Hence the resistance of the circuit is less and so the current through L_1 is greater and it is brighter. The pd across L_1 is also greater so the pd across L_2 is less, so the current is less and so L_2 is dimmer. L_3 has the same brightness as L_2 because they are identical and have the same current.

Q2 (a) For each coulomb of charge which flows through the two resistors, 5 J of energy is transferred in the left-hand resistor and 3 J in the right. So the total energy transfer per coulomb is 8 J. This must also be the energy transferred in the battery so the pd is 8 V.

(b) 8 V.

Q3 $R = 12\ \Omega \longrightarrow I = 0.50\ A$ ⎯▭⎯ $R = 24\ \Omega \longrightarrow I = 0.25\ A$ ⎯▭─▭⎯

$R = 36\ \Omega \longrightarrow I = 0.17\ A$ ⎯▭─▭─▭⎯

$R = 6.0\ \Omega \longrightarrow I = 1.0\ A$ $R = 4.0\ \Omega \longrightarrow I = 1.50\ A$

$R = 18\ \Omega \longrightarrow I = 0.33\ A$

$R = 8.0\ \Omega \longrightarrow I = 0.75\ A$

Q4 (a) Current in 12 Ω resistor = 0.30 A $\times \dfrac{20\ \Omega}{12\ \Omega}$ = 0.50 A

∴ Current in 15 Ω resistor = 0.30 A + 0.50 A = 0.80 A

$P = I^2 R,$

so total power = $(0.80)^2 \times 15 + (0.30)^2 \times 20 + (0.50)^2 \times 12 = 14.4\ \Omega$

(b) The resistance of the 15 Ω / X combination is less than that of the 15 Ω alone, so the total resistance of the circuit decreases, so the total current increases. Hence the pd across the 20 Ω / 12 Ω combination increases and so the pd across the 15 Ω decreases.

Q5 (a) The top component in the circuit is an LDR. As the light level increases, the resistance of the LDR decreases. Hence, the total resistance of the circuit decreases and so the current increases. The pd across the fixed resistor, which is equal to V_{OUT}, increases.

(b) At 37°C, resistance of thermistor = 6.4 kΩ.

The thermistor needs to be in the bottom position because its resistance increases as the temperature decreases, thus increasing the output voltage.

Using potential divider to calculate the resistance, R, of the series resistor:

$5\,V = \dfrac{6.4\,k\Omega}{R+6.4\,k\Omega} \times 12\,V$ ∴ $R = 9.0\,k\Omega$ (2 sf)

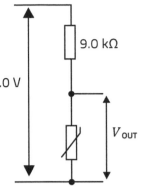

Q6 (a) 9.0 J of energy is transferred from chemical store per coulomb of charge passing through the battery.

(b) (i) Chemical energy transferred per second = 9.0 V × 1.5 A = 13.5 W.

Energy transferred by the electric current in the external circuit = 7.8 V × 1.5 A = 11.7 W

Energy transferred to thermal energy by the electric current inside the battery is the difference in these, i.e. 13.5 W – 11.7 W = 1.8 W

(ii) $V = E - Ir$ So $r = \dfrac{E-V}{I} = \dfrac{9.0\,V - 7.8\,V}{1.5\,A} = 0.8\,\Omega$ [or, use $P_r = I^2r$, with $P_r = 1.8$ W and $I = 1.5$ A]

Q7 With one 10 Ω resistor, the pd is 6.5 V, so $I = \dfrac{6.5\,V}{10\,\Omega} = 0.65$ A

Applying $V = E - Ir \longrightarrow 6.5 = E - 0.65r$ (1)

With two 10 Ω resistors, the total external resistance is 5 Ω, and the pd is 6.0 V

so $I = \dfrac{6.0\,V}{5.0\,\Omega} = 1.2\,A \longrightarrow 6.0 = E - 1.2r$ (2)

Subtracting equation (2) from equation (1) $\longrightarrow 0.5 = 0.55r$

∴ $r = 0.91\,\Omega$

Substituting in (1) and rearranging $\longrightarrow E = 6.5 + 0.65 \times 0.91 = 7.1$ V

Q8 $V = E \times \dfrac{R}{R+r}$ and $I = \dfrac{E}{R+r}$; when $R = r$, $V = \dfrac{E}{2}$ and $I = \dfrac{E}{2r}$, so $P_{out} = \dfrac{E^2}{4r}$

For the conventional cell: $P_{out} = \dfrac{(1.5)^2}{4 \times 0.3} = 1.875$ W

For the Ni-Cd cell, $P_{out} = \dfrac{(1.2)^2}{4 \times 0.035} = 10.3$ W = 5.5 × the max power from the conventional cell.

Q9 (a) 18 Ω resistor in parallel with a series combination of 3.3 Ω and 10 Ω.

$R = \dfrac{18 \times (3.3 + 10)}{18 + (3.3 + 10)}\Omega = \dfrac{239.4}{31.3}\Omega = 7.6\,\Omega$ (2 sf)

(b) Eliminating V from the equations: $E - Ir = IR$

∴ $E = I(R + r)$

Dividing by $EI \longrightarrow \dfrac{1}{I} = \dfrac{R}{E} + \dfrac{r}{E}$

So a graph of $1/I$ against R is a straight line of gradient $1/E$ and intercept on the $1/I$ axis of r/E.

(c) 2.3   2.9   3.2   3.7   4.3   5.3

(d)

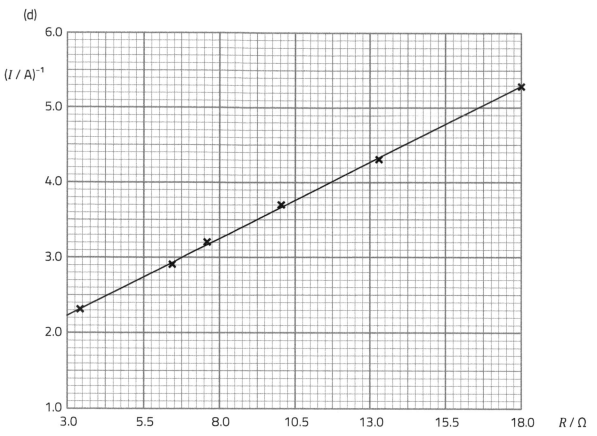

(e) Gradient = $\dfrac{5.3 - 2.2}{18.0 - 3.0}$ = 0.207 V⁻¹, giving a value of E of $\dfrac{1}{0.207\ \text{V}^{-1}}$ =4.84 V

Applying $y = mx + c$, then $c = y - mx$. Using the values (10.5, 3.76) gives c = 1.59 A⁻¹.

Then $r = \dfrac{\text{intercept}}{\text{gradient}} = \dfrac{1.59\ \text{A}^{-1}}{0.207\ \text{V}^{-1}}$ = 7.7 Ω.

This shows that the teacher's value of emf is consistent but the internal resistance is higher than the student's results.

Q10 (a) pd across the terminals of the battery = 3 × 1.2 V = 3.6 V

Current = $\dfrac{1.5\ \text{W}}{2.5\ \text{V}}$ = 0.6 A. pd across resistor = 3.6 V − 2.5 V = 1.1 V

∴ Resistance of resistor, $R = \dfrac{1.1\ \text{V}}{0.6\ \text{A}}$ = 1.8 Ω

(b) Fraction = $\dfrac{I^2 R}{I^2 R + I^2 R_{\text{lamp}}} = \dfrac{R}{R + R_{\text{lamp}}} = \dfrac{1.8\ \Omega}{1.8\ \Omega + (2.5/0.6)\ \Omega}$ = 0.30

Q11 (a) V_{total} = 0.70 V + 0.03 A × 820 Ω = 25.3 V

(b) Resistance needed for 10 mA = = 790 Ω

Resistance needed for 25 mA = $\dfrac{(9.0 - 1.9)\text{V}}{0.025\ \text{A}}$ = 316 Ω

So 470 Ω and 680 Ω are suitable

Section 2.4: Capacitance

Q1 $Q = CV$ = 22 mF × 12 V = 260 mC (2 sf)

So the charges on the plates are +260 mC and −260 mC.

Q2 (a) $C = \dfrac{\varepsilon_0 A}{d}$, so $d = \dfrac{\varepsilon_0 A}{C} = \dfrac{8.85 \times 10^{-12}\ \text{F m}^{-1} \times (0.10\ \text{m})^2}{500 \times 10^{-12}\ \text{F}}$ = 1.8 × 10⁻⁴ m (0.18 mm)

(b) Putting a polymer between the plates of a capacitor increases its capacitance. The capacitance is inversely proportional to the plate separation. Hence, to achieve the same capacitance, the plate separation needs to be bigger.

Q3 (a) $C = \dfrac{\varepsilon_0 A}{d} = \dfrac{8.85 \times 10^{-12}\, \text{F m}^{-1} \times 64 \times 10^{-4}\, \text{m}^2}{0.40 \times 10^{-3}\, \text{m}} = 1.42 \times 10^{-10}\, \text{F}$

$Q = CV = 1.42 \times 10^{-10}\, \text{F} \times 30\, \text{V} = 4.2 \times 10^{-9}\, \text{C}$ (2 sf)

(b) $U = \frac{1}{2} CV^2 = 0.5 \times 1.42 \times 10^{-10}\, \text{F} \times (30\, \text{V})^2 = 6.4 \times 10^{-8}\, \text{J}$

(c) $E = \dfrac{V}{d} = \dfrac{30\, \text{V}}{0.40 \times 10^{-3}\, \text{m}} = 75\,000\, \text{V m}^{-1}$

Q4 (a) $U = \frac{1}{2}\dfrac{Q^2}{C}$. As Q is constant $U \propto C^{-1}$.

But $C = \dfrac{\varepsilon_0 A}{d}$, so $C \propto d^{-1}$ and so $U \propto d$. Hence the energy is doubled.

(b) The opposite charges attract, so work has to be done to separate them. The energy has come from the agency which separated the plates.

Q5 $U = \frac{1}{2}CV^2 = \frac{1}{2}\dfrac{\varepsilon_0 A}{d}V^2$

But $E = \dfrac{V}{d}$, so $V = Ed$ and substituting for V gives:

$U = \frac{1}{2}\dfrac{\varepsilon_0 A}{d}(Ed)^2$ which simplifies to $U = \frac{1}{2}\varepsilon_0 E^2 \times (Ad) = \frac{1}{2}\varepsilon_0 E^2 \times$ volume

Q6 Total capacitance, $C = \dfrac{C_1 C_2}{C_1 + C_2} = \dfrac{7.0 \times 3.0}{7.0 + 3.0}\, \mu\text{F} = 2.1\, \mu\text{F}$

So charge flow in charging, $Q = CV = 2.1\, \mu\text{F} \times 20\, \text{V} = 42\, \mu\text{C}$

So the charges on the plates (from left to right) are: $-42\, \mu\text{C}, +42\, \mu\text{C}, -42\, \mu\text{C}, +42\, \mu\text{C}$

Q7 (a) Capacitance of series combination $= \dfrac{120 \times 40}{120 + 40} = 30\, \mu\text{F}$

So total capacitance $= 30\, \mu\text{F} + 30\, \mu\text{F} = 60\, \mu\text{F}$

(b) (i) $Q = CV = 60\, \mu\text{F} \times 12\, \text{V} = 720\, \mu\text{C}$

(ii) **Either**: Capacitors in series in ratio $3:1$
So pds in ratio $1:3$, i.e. $\frac{1}{4}$ on the 120 μF
So the pd across the 120 μF $= 3\, \text{V}$
Or: Charge on the series combination $= \frac{1}{2} \times 720\, \mu\text{C} = 360\, \mu\text{C}$
So charge on 120 μF $= 360\, \mu\text{C}$

So pd on 120 μF $= \dfrac{Q}{C} = \dfrac{360\, \mu\text{C}}{120\, \mu\text{F}} = 3\, \text{V}$

Q8 (a) The total separated charge on the capacitors is unchanged but it is now shared between them. Because the capacitors have equal value and the pds must be equal, they share the charge equally. So each had half the charge and the pd is half, i.e. 4.5 V.

(b) Initial energy on $C_1 = \frac{1}{2}CV^2 = 0.5 \times 50\, \text{mF} \times (9.0\, \text{V})^2 = 2.025\, \text{J}$ [2.0 J to 2 sf]
Final energy $= 2 \times (0.5 \times 50\, \text{mF} \times (4.5\, \text{V})^2) = 1.0125\, \text{J}$ [1.0 J to 2 sf]
So energy change $= 1.0\, \text{J} - 2.0\, \text{J} = -1.0\, \text{J}$ [i.e. a 'loss' of 1.0 J]

Q9 (a) The final (maximum) charge. [CV_0 would be an acceptable answer.]

(b) (i) $Q = CV$. Substituting in $Q = Q_0(1 - e^{-t/RC})$ for Q and Q_0
$\longrightarrow CV_c = CV_0(1 - e^{-t/RC})$ and dividing by C gives the required equation.

(ii) $V_R = V_0 - V_c = V_0 - V_0\left(1 - e^{-t/RC}\right) = V_0 e^{-t/RC}$

(iii) $V_R = V_0 e^{-t/RC}$. Dividing by $R \longrightarrow \dfrac{V_R}{R} = \dfrac{V_0}{R}e^{-t/RC}$

$I = \dfrac{V_R}{R}$. When the capacitor is uncharged, $V_R = V_0$, so $I_0 = \dfrac{V_0}{R}$

Hence $I = I_0 e^{-t/RC}$

(c) (i) Capacitance, C, is defined by the equation $C = \frac{Q}{V}$, $[C] =$ A s V^{-1}

Resistance, R, is defined by the equation $R = \frac{V}{I}$, $\therefore [R] =$ V A^{-1}

$\therefore [RC] =$ V A^{-1} × A s V$^{-1} =$ s. $\therefore \left[\frac{Q_0}{RC}\right] = $ C s$^{-1} =$ ampère

(ii) $I_0 = $ gradient of tangent at $t = 0$.

$\therefore$ Gradient $= \frac{V_0}{R} = V_0 \times \frac{1}{R} = \frac{Q_0}{C} \times \frac{1}{R} = \frac{Q_0}{RC}$

Q10 (a)

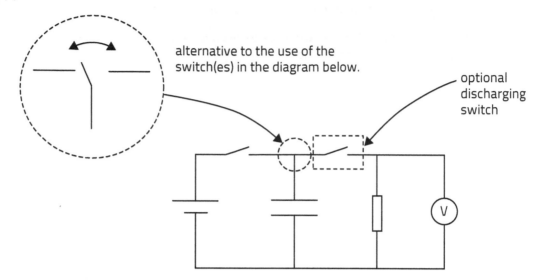

alternative to the use of the switch(es) in the diagram below.

optional discharging switch

(b) (i) Time for V_c to halve from 4.0 V to 2.0 V = 24.0 s – 8.5 s = 15.5 s
Time for V_c to halve from 2.0 V to 1.0 V = 39.5 s – 24.0 s = 15.5 s
Thus the half-life is constant, showing an exponential decay.

(ii) [Easy method]: After one time constant, the value of V_c falls to $1/e = 0.37$ of the original value.
0.37×5.9 V $= 2.2$ V $\longrightarrow$ time = 22 s.
[More difficult, but equally valid]: $V_c = V_0 e^{-t/RC}$, so $\ln\left(\frac{V_0}{V_c}\right) = \frac{t}{RC}$

Choose, e.g. $t = 39$ s $\longrightarrow V_c = 1.0$ V, so $\ln\left(\frac{5.9}{1.0}\right) = \frac{39}{RC} \longrightarrow RC$ (time constant) = 22 s

(c) $V_c = V_0 e^{-t/RC}$: taking logs $\longrightarrow \ln (V_0/\text{volt}) = \ln (V_c/\text{volt}) - \frac{t}{RC}$

$\therefore$ Intercept $= \ln (V_c / \text{volt}) = \ln (5.9$ V $/$ V$) = \ln 5.9 = 1.8$ (2 sf)
and gradient $= -(RC)^{-1} = -1/22 = -0.045$ (2 sf)

Section 2.5: Solids under stress

Q1 (a) $F = $ tension (or force); $k = $ spring constant; $x = $ extension

(b) $k = \frac{F}{x}$, so $[k] = \frac{\text{kg m s}^{-2}}{\text{m}} = $ kg s^{-2}

(c) The extension is within the elastic limit.

Q2 Crystalline: one in which the atoms are arranged in a regular array, with long-range order, e.g. diamond.
Amorphous: one in which the arrangement of the atoms has no long-range order, e.g. glass.
Polymeric: one with long chain molecules consisting of multiple repeat units, e.g. polythene.

Q3 (a)

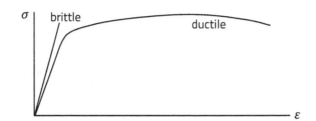

(b) [One answer] Concrete can be made stronger under tension by pre-compressing it. Metal rods under tension are placed in the concrete. When the tension is released, the concrete is placed under compression. Hence any cracks cannot propagate.

Q4 (a) Using $mg = kx$, $k = \dfrac{mg}{x} = \dfrac{0.300 \text{ kg} \times 9.81 \text{ N kg}^{-1}}{0.151 \text{ m}} = 19.49 \text{ N m}^{-1}$

Fractional uncertainties: $p_m = \dfrac{6}{300} = 0.020$; $p_x = \dfrac{0.2}{15.1} = 0.013$

$\therefore p_k = 0.020 + 0.013 = 0.033$

Absolute uncertainty: $\Delta k = 0.033 \times 19.49 = 0.6$ (1 sf)

$\therefore k = (19.5 \pm 0.6) \text{ N m}^{-1}$

(b) Method: 1. Pull the mass (300 g) down a few cm and release it.

2. Use a stop watch to measure the time for (say) 20 oscillations.

3. Repeat 2 several times, calculate the mean value and estimate the uncertainty.

4. Calculate the period and the uncertainty by dividing by the number of oscillations.

5. Calculate T and its uncertainty by putting m and k in the equation and comparing with the measured value of the period..

Q5 (a) A ductile material is one that can be drawn into a wire (or can be deformed plastically without becoming brittle).

(b) (i) Consisting of a large number of interlocking crystals.

(ii) The diagram shows planes of atoms in a crystal. There is an extra half plane (ending at Y). This kind of defect is called an edge dislocation. When the tensile force, F, is applied, the bonds at **X** break and new ones are formed with **Y**. The dislocation moves to the right and causes a change in shape that does not reverse when the tension is removed.

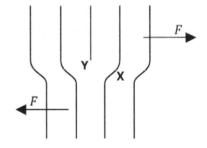

Q6 Elastic strain is a deformation which is reversed when the applied force is removed. Stress is the tension divided by the cross-sectional area of the material.

The elastic limit is the value of stress below which the material exhibits elastic strain only and above which it exhibits plastic strain.

Plastic strain is strain which is not reversed when the applied force is removed.

Q7

$\dfrac{F_A}{F_B} = 1$	$\dfrac{\varepsilon_A}{\varepsilon_B} = 6$	$\dfrac{W_A}{W_B} = 3$
$\dfrac{\sigma_A}{\sigma_B} = 4$	$\dfrac{\Delta l_A}{\Delta l_B} = 3$	$\dfrac{W_A/V_A}{W_B/V_B} = 24$

Calculation

The tension is the same throughout, so $F_A = F_B$. Area of B = 4 × area of A, so $\sigma_A = 4\sigma_B$.

$\varepsilon_A = \dfrac{\sigma_A}{E_A} = \dfrac{4\sigma_B}{E_B/1.5} = 6\varepsilon_B \therefore \dfrac{\varepsilon_A}{\varepsilon_B} = 6$; $\Delta l_A = \varepsilon_A l_A = 6\varepsilon_B \times \frac{1}{2}l_B = 3\Delta l_B$.

$W_A = F_A \Delta l_A = F_B \times 3\Delta l_B = 3W_B$; $\dfrac{W_A}{V_A} = \dfrac{3W_B}{\frac{1}{8}V_B} = 24\dfrac{W_B}{V_B}$

Q8 (a) To remove kinks from the wire

(b) $E = \dfrac{\sigma}{\varepsilon} = \dfrac{Fl_0}{A\Delta l} = \dfrac{4Fl_0}{\pi D^2 \Delta l}$

$\Delta D = \dfrac{D_{max} - D_{min}}{2} = \dfrac{0.25 - 0.23}{2} = 0.01$; so ∴ % uncertainty in $D = \dfrac{0.01}{0.24} \times 100\% = 4.2\%$

∴ $p(A) = 2 \times 4.2\% = 8.4\%$; $p(\Delta l) = \dfrac{0.1}{3.1} \times 100\% = 3.2\%$

The uncertainties in length and the force are negligible, so the percentage uncertainty in $E = 8.4\% + 3.2\% = 11.6\% = 10\%$ (1 sf) or 12% (2 sf)

(c) If the D is (say) halved, A is divided by 4 and so σ is multiplied by 4, so Δl is multiplied by 4. $p(A)$ is × 2 and $p(\Delta l)$ is ÷ 4. In this case $p(E)$ would be 16.8% + 0.8% ~ 18% which is a greater uncertainty.

Q9 (a) The stress at which large plastic strain occurs (or at which edge dislocations move).

(b) $\sigma = 60$ MPa ∴ $\varepsilon = \dfrac{\sigma}{E} = \dfrac{60 \text{ MPa}}{2.00 \text{ GPa}} = 3.0 \times 10^{-4}$

EPE $= \frac{1}{2} Fx = \frac{1}{2} \times 60 \times 10^6$ Pa $\times \pi (0.015 \text{ m})^2 \times 3.0 \times 10^{-4} \times 5\,000$ m

$= 3.2 \times 10^4$ J

Q10 (a) Work done $= \frac{1}{2} F\Delta x = \frac{1}{2} \times 280$ N $\times 0.76$ m $= 106.4$ J $= 110$ J (2 sf)

(b) KE given to arrow $= 90\% \times 106.4$ J $= 95.76$ J

∴ velocity $= \sqrt{\dfrac{2E_k}{m}} = \sqrt{\dfrac{2 \times 95.76 \text{ J}}{0.050 \text{ kg}}} = 61.9$ m s^{-1}

Vertical component (= horizontal component) of velocity $= 61.9 \cos 45° = 43.8$ m s^{-1}

Time in air $= \dfrac{v - u}{a} = \dfrac{43.8 - (-43.8)}{9.81} = 8.92$ s.

∴ Range (using horizontal velocity) $= 43.8$ m s$^{-1} \times 8.92$ s $= 390$ m (2 sf), so the claim is slightly exaggerated (unless there is some aerodynamic lift).

Q11 All the kinetic energy of the train needs to be transferred to the elastic energy in the springs before the springs reach maximum compression, i.e. $E_k = \frac{1}{2} Fx = \frac{1}{2} kx^2$. The smaller the spring constant, the larger the distance needed to stop the train. A large distance needs a long housing for the spring which is a disadvantage.

However, $F = \sqrt{\dfrac{2E_k}{x}}$, so this longer decelerating distance results in a lower maximum force, so a smaller deceleration of the train, i.e. a less violent halt, which is an advantage.

Section 2.6: Electrostatic and gravitational fields of force

Q1

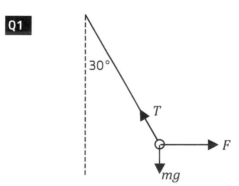

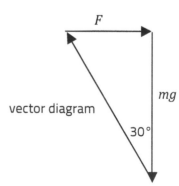

vector diagram

(a) From the vector diagram, $\tan 30° = \dfrac{F}{mg} = \dfrac{F}{2.00 \times 10^{-7} \times 9.81}$ ∴ $F = 1.13 \times 10^{-6}$ N » 1.1 × 10^{-6} N

(b) Separation of charges = 0.20 m. Using $F = \dfrac{1}{4\pi\varepsilon_0} \dfrac{Q_1 Q_2}{r^2}$

∴ $1.13 \times 10^{-6} = 9.0 \times 10^9 \times \dfrac{Q^2}{(0.2)^2}$

∴ $Q = 2.2 \times 10^{-9}$ C (2 sf)

Q2 (a) Assuming that the Moon is spherically symmetric, the field strength at its surface is given by:

$$g = \frac{GM}{r^2}, \text{ so } M = \frac{gr^2}{G} = \frac{1.62\,\text{N kg}^{-1} \times (1737 \times 10^3\,\text{m})^2}{6.67 \times 10^{-11}\,\text{N m}^2\,\text{kg}^{-2}} = 7.33 \times 10^{22}\,\text{kg}$$

(b) $F = \dfrac{GM_1 M_2}{r^2} = \dfrac{6.67 \times 10^{-11} \times 5.97 \times 10^{24} \times 7.33 \times 10^{22}}{(3.84 \times 10^8)^2}\,\text{N} = 1.98 \times 10^{20}\,\text{N}$

Q3 (a) $F_G = \dfrac{GM^2}{r^2}$; $F_E = \dfrac{1}{4\pi\varepsilon_0}\dfrac{Q^2}{r^2}$,

so $\dfrac{F_E}{F_G} = \dfrac{1}{4\pi\varepsilon_0}\dfrac{Q^2}{GM^2} = 9.0 \times 10^9 \dfrac{(1.60 \times 10^{-19})^2}{6.67 \times 10^{-11} \times (1.67 \times 10^{-27})^2} = 1.24 \times 10^{36}$ (3 sf)

(b) There are almost exactly the same number of protons and electrons on each of the Earth and the Sun, so these bodies are virtually electrically neutral. Hence the electrostatic force between them is negligible. The masses of the protons and electrons are both positive (there are no negative masses) so these both contribute positively to the gravitational force.

Q4 (a)

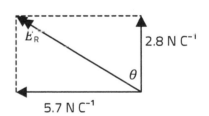

(b) P is equidistant from the two charges, so the magnitudes of the fields due to the charges are equal. The fields are directed radially towards the − charge and away from the + charge, so from the symmetry, the vertical components cancel, the horizontal components add and the resultant field is horizontal to the right.

Q5 (a) (i) $E = \dfrac{1}{4\pi\varepsilon_0}\dfrac{Q}{r^2} = 9.0 \times 10^9 \times \dfrac{(-3.0 \times 10^{-12})}{(0.12 \cos 35°)^2}$ South $= -2.79$ (N C^{-1}) or (V m^{-1})

i.e. 2.8 N C^{-1} (2 sf) due North.

(ii) $E = 9.0 \times 10^9 \times \dfrac{3.0 \times 10^{-12}}{(0.12 \sin 35°)^2}$ West $= 5.7$ N C^{-1} due West.

(iii) $E_R = \sqrt{(2.8)^2 + (5.7)^2} = 6.4$ N C^{-1}

$\theta = \tan^{-1}\left(\dfrac{5.7}{2.8}\right) = 64°$. ∴ Bearing $= 360° - 64°$

∴ 6.4 N C^{-1} at 296°

(b) (i) $V = \dfrac{1}{4\pi\varepsilon_0}\dfrac{Q}{r}$, ∴ $V_{\text{tot}} = 9.0 \times 10^9 \times 3.0 \times 10^{-12} \times \left(-\dfrac{1}{0.12 \cos 35°} + \dfrac{1}{0.12 \sin 35°}\right)$

$= 0.118$ V

(ii) Initial potential energy of proton $= eV = 1.60 \times 10^{-19}$ C $\times 0.118$ V $= 1.89 \times 10^{-20}$ J

∴ Max kinetic energy $= 1.89 \times 10^{-20}$ J (when all PE has been lost)

∴ $v = \sqrt{\dfrac{2E_k}{m_p}} = \sqrt{\dfrac{2 \times 1.89 \times 10^{-20}\,\text{J}}{1.67 \times 10^{-27}\,\text{kg}}} = 4760$ m s^{-1} (3 sf)

Q6 (a) **Either** The equipotential surfaces are at right angles to the field lines, which are radial.

Or The potential is given by: $V = \dfrac{1}{4\pi\varepsilon_0}\dfrac{Q}{r}$, so all points at the same radius, r, have the same potential.

(b) The potential is inversely proportional to the distance: $V = \dfrac{1}{4\pi\varepsilon_0}\dfrac{Q}{r}$, so $r \propto \dfrac{1}{V}$.

$\dfrac{1}{2} - \dfrac{1}{3} = \dfrac{1}{6}$ but $\dfrac{1}{3} - \dfrac{1}{4} = \dfrac{1}{12}$, which is only half as much.

Q7 (a) The defined zero of gravitational potential energy is infinity. Gravity is always an attractive force, hence work must be done to separate objects to infinity. Doing work on a system transfers energy to the system; hence for separations less than infinite (!) the potential energy (and therefore the potential at a non-infinite point) must be negative.
[Note: In an exam, the first two sentences in this answer would probably score you both marks (and there is not really enough room to write more) but the last sentence is needed for a full explanation.]

(b) (i) The equation $\Delta(PE) = mgh$ only applies in situations in which g may be taken as constant. In fact, g varies as r^{-2} so, for large increases in r, this must be taken into account.

(ii) Let mass of rocket be m. Then, applying conservation of energy:
Initial KE + Initial PE = PE at greatest height [because KE = 0 at greatest height].

$$\tfrac{1}{2}m \times (3000)^2 - \frac{6.67 \times 10^{-11} \times 6.42 \times 10^{23}\,m}{3390 \times 10^3} = -\frac{6.67 \times 10^{-11} \times 6.42 \times 10^{23}\,m}{r}$$

$$\therefore -8.13 \times 10^6 = -\frac{4.28 \times 10^{13}}{r}, \therefore r = \frac{4.28 \times 10^{13}}{8.13 \times 10^6}\ m = 5267\ km$$

$\therefore$ Height = 5267 – 3390 = 1900 km (2 sf)

Q8 (a) (i)

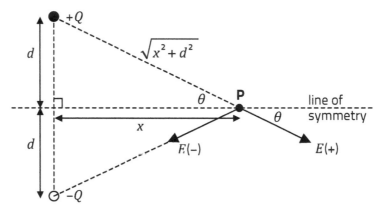

Magnitudes: $E(+) = E(-) = \dfrac{1}{4\pi\varepsilon_0}\dfrac{Q}{x^2+d^2}$. The horizontal components of $E(+)$ and $E(-)$ are equal and opposite, to cancel. The vertical components add.

$$\therefore E = \frac{1}{4\pi\varepsilon_0}\frac{2Q}{x^2+d^2}\sin\theta = \frac{1}{4\pi\varepsilon_0}\frac{2Q}{x^2+d^2}\frac{d}{\sqrt{x^2+d^2}} = \frac{1}{2\pi\varepsilon_0}\frac{Qd}{(x^2+d^2)^{3/2}}$$

The direction of E is vertically downwards in the diagram.

(ii) The <u>resultant</u> electric field strength is always at right angles to the line of symmetry, so there is no component of the electric force on a test charge in the direction of motion (along the symmetry line) and so the work done is zero. Hence, Adam is right and Bethan is wrong.
Alternative answer: Work has to be done against the repulsion of $+Q$ in bringing a (positive) test charge from infinity. However, this is balanced by the equal negative work needed to bring the test charge against the attraction of $-Q$ and so the net work is zero. Adam is right and Bethan is wrong.

(b) For the +/– charge combination, the electric field strength, E, is always at right angles to the line of symmetry; its maximum value is when $x = 0$ and it decreases towards zero as $x \longrightarrow \infty$.
For two equal positive charges the direction of E is always along the line of symmetry, in the $+x$ direction. E is zero when $x = 0$ (because the fields from the two charges are equal and opposite), rises to a maximum as x increases before decreasing towards zero as $x \longrightarrow \infty$ (inverse square at large distances).

Section 2.7: Using radiation to investigate stars

Q1 Power $= \dfrac{\text{energy transfer}}{\text{time}}$, $\therefore$ W $=$ J s^{-1}.

$\therefore$ W m^{-2} $=$ (kg m^2 s^{-2}) s^{-1} m^{-2}

$\qquad = $ kg s^{-3}

Q2 A black body is one which absorbs all electromagnetic radiation which is incident upon it. No body emits more radiation [due to its temperature] at any wavelength than a black body.

Q3 $L = \sigma A T^4 = \sigma(4\pi r^2)T^4$,

$\therefore \dfrac{L_{\text{red dwarf}}}{L_{\text{Sun}}} = \left(\dfrac{r_{\text{red dwarf}}}{r_{\text{Sun}}}\right)^2 \left(\dfrac{T_{\text{red dwarf}}}{T_{\text{Sun}}}\right)^4$

$\therefore L_{\text{red dwarf}} = \left(\dfrac{1}{4}\right)^2 \left(\dfrac{1}{2}\right)^4 \times 4 \times 10^{26}$ W $= \dfrac{4 \times 10^{26} \text{ W}}{256} = 1.6 \times 10^{24}$ W

Q4 (a) $L = \left(\dfrac{0.7M_\odot}{M_\odot}\right)^4 L_\odot = 0.24\ L_\odot$

(b) For a number $x < 1$, the value of x^n decreases as n increases. So, in fact, n should be greater than 4, and Alex is incorrect.

[Note: $(0.7)^5 = 0.17$, which is closer to the actual value.]

Q5 Assuming that the star emits as a black body, the wavelength, λ_{max}, of the peak spectral intensity relates to the kelvin temperature, T, by Wien's displacement law: $\lambda_{max} = \dfrac{W}{T}$, where W is a constant, 2.9×10^{-3} m K. Hence the temperature of the star's photosphere (its surface) can be determined. The intensity, I, of the radiation from the star can be determined from the total area under the graph. The luminosity, L, of the star can then be calculated using $L = 4\pi d^2 \times I$, where d is the distance to the star.

The luminosity, L is given by the Stefan–Boltzmann law: $L = A\sigma T^4$, where A is the surface area of the star. Hence, A can be calculated (knowing T from Wien's law) and thus the radius, r, from $A = 4\pi r^2$.

Q6 The continuous emission spectrum is the radiation which is given out at all wavelengths. The line absorption spectrum is the missing radiation at particular wavelengths in the spectrum.

Q7 Atoms in the star's atmosphere can be promoted to higher states by the absorption of radiation with a photon energy equal to the difference in energy levels. This results in the dark lines (Fraunhofer lines) which are characteristic of the elements of the atoms in the star.

Q8 (a) $T = \dfrac{W}{\lambda_{max}} = \dfrac{2.90 \times 10^{-3}\text{m K}}{501\text{ nm}} = 5790$ K

(b) If the Sun were a black body of temperature 5790 K, its luminosity would be:

$L = 4\pi\left(\dfrac{d}{2}\right)^2 \sigma T^4 = 4\pi \times \left(\dfrac{1.39 \times 10^9 \text{ m}}{2}\right)^2 \times (5.67 \times 10^{-8}\text{ W m}^{-2}\text{ K}^{-4}) \times (5790\,\text{K})^4$

$\qquad = 3.87 \times 10^{26}$ W

This is very close to the website value so it is consistent with the Sun emitting as a black body.

Q9 (a) (i) $\lambda_{max} = \dfrac{W}{T} = \dfrac{2.90 \times 10^{-3}\text{m K}}{4\,000\text{ K}} = 725$ nm

(ii) Near infra-red (very close to the red end of the visible spectrum).

(b) The ratio of the e-m power emitted per unit area from the sunspot compared to that of the Sun's photosphere is given by $\left(\dfrac{4000}{6000}\right)^4 = 0.20$ (2 sf). Also, nearly all the radiation is emitted in the infra-red – the Sun emits mainly in the visible – so any visible image will contain very little radiation from the sunspot.

Q10 Photon energy $E_{ph} = hf = \dfrac{hc}{\lambda}$

Wien's law: $\lambda_{max} = \dfrac{W}{T}$, photon energy $E_{ph\,max} = \dfrac{hc}{\lambda_{max}} = \dfrac{hc}{W}T$. So Bryn is correct.

Q11 Multiwavelength astronomy is taking images of objects in the universe using several regions of the electromagnetic spectrum, e.g. X-rays, visible, radio. The different regions give information about different conditions and processes that occur, e.g. comparing UV, visible and infra-red images of stars enables us to compare their temperatures.

Q12 $\lambda_{max} = \frac{W}{T}$. For cosmic background radiation, $\lambda_{max} \sim 1$ mm and galactic molecular clouds have $\lambda_{max} \sim 0.1$ mm, so using microwave and submillimetre radiation gives us information about processes there.

λ_{max} for stars like the Sun (6000 K) $\sim$ 500 nm, so the range for Red Giant to Blue Giant stars is ~ 1 µm $-$ 50 nm, which spans the IR and UV regions of the e-m spectrum.

Supernovae, black holes and inter-galactic gas have $\lambda_{max} \sim 10^{-9} - 10^{-11}$ m, which is in the X-ray region of the spectrum, so X-ray astronomy can reveal processes in these objects.

Q13 $E_{ph} = hf = \frac{hc}{\lambda}$

For 10 nm $E_{ph} = \frac{6.63 \times 10^{-34} \times 3.00 \times 10^8}{10 \times 10^{-9}} = 2.0 \times 10^{-17}$ J $= \frac{2.0 \times 10^{-17} \text{J}}{1.60 \times 10^{-19} \text{J/eV}} = 120$ eV

For 400 nm, the values are 1/40 of these, i.e. 5.0×10^{-19} J, 3.0 eV

Q14 (a) Photon energy $= \frac{6.63 \times 10^{-34} \times 3.00 \times 10^8}{1.0 \times 10^{-6} \times 1.60 \times 10^{-19}} = 1.2$ eV . If there are He$^+$ ions in the 3rd excited state (energy -3.4 eV) they can absorb photons of energy 1.2 eV and enter the 4th excited state (energy -2.2 eV) because the difference in energy levels is 1.2 eV. The light in the direction of the emission from the star is depleted in photons of this energy, giving rise to the dark line.

(b) Transitions from -6.0 eV $\longrightarrow -3.4$ eV require 2.6 eV and $-3.4 \longrightarrow -1.5$ require 1.9 eV, both of which are in the visible range. (Note: Ionisation from the -2.2 eV level will also absorb visible photons but this will not be a single line because there is no single upper energy level.)

(c) In order to produce these lines, the He$^+$ ions must first be excited to the -6.0 eV energy level, which requires an energy of 48.4 eV. Photons in the visible range have energies around $2-3$ eV and the Sun's spectrum (from its 6 000 K surface) only extends a small way into the UV, so there is not enough energy to do this and Eleri is correct.

Section 2.8: Orbits and the wider universe

Q1 Orbital period = 1 year = 60 × 60 × 24 × 365.25 s = 3.156×10^7 s

Orbital radius = 150 × 10^6 km = 1.50×10^{11} m

$T = 2\pi \sqrt{\frac{r^3}{GM_\odot}}$, $\therefore M_\odot = \frac{4\pi^2}{T^2} \times \frac{r^3}{G} = \frac{4\pi^2}{(3.156 \times 10^7)^2} \times \frac{(1.50 \times 10^{11})^3}{6.67 \times 10^{-11}} = 2.0 \times 10^{30}$ kg

Q2 (a) $g = \frac{GM}{r^2}$, $\therefore M = \frac{gr^2}{G} = \frac{9.81 \times (6.37 \times 10^6)^2}{6.67 \times 10^{-11}} = 5.97 \times 10^{24}$ kg

(b) The mass distribution is spherically symmetric.

(c) $\rho = \frac{M}{V} = \frac{5.97 \times 10^{24} \text{kg}}{\frac{4}{3}\pi \times (6.37 \times 10^6 \text{m})^3} = 5510$ kg m^{-3}

Q3 The position of the star appears reasonable. It should be at one of the foci of the ellipse.

Given that S is correct, then X is clearly incorrect. It is the closest distance of approach between the planet and the star. Hence, this is the point at which the planet moves most quickly. It should be at the opposite end of the major axis [or S should be at the right-hand focus].

Q4 For both parts: 1 year = 60 × 60 × 24 × 365.25 s = 31 557 600 s [31 536 000 if 365 days used]

(a) $T = 2\pi \sqrt{\frac{d^3}{GM}}$, so $d^3 = \frac{GMT^2}{4\pi^2} = \frac{6.67 \times 10^{-11} \times 1.99 \times 10^{30} \times (3.156 \times 10^7)^2}{4\pi^2}$

$\therefore d = 1$ AU $= 1.496 \times 10^{11}$ m = 150 million km (3 sf) [149 million km if 365 days used]

(b) Distance = vt = 3.00×10^5 km s^{-1} × 3.156×10^7 s = 9.47×10^{12} km

[9.46×10^{12} km if 365 days used]

Q5 (a) Period = 27.3 day = 60 × 60 × 24 × 27.3 s = 2.36×10^6 s

$$a = r\omega^2 = r\left(\frac{2\pi}{T}\right)^2 = 3.83 \times 10^8 \text{ m} \times \left(\frac{2\pi}{2.36 \times 10^6 \text{s}}\right)^2 = 2.72 \times 10^{-3} \text{ m s}^{-2}$$

(b) (i) ratio = $\dfrac{9.81}{2.72 \times 10^{-3}}$ = 3610

   (ii) ratio = $\left(\dfrac{3.83 \times 10^8 \text{ m}}{6.37 \times 10^6 \text{ m}}\right)^2$ = 3620

(c) At radius r_1: $g_1 = \dfrac{GM}{r_1^2}$. At r_2: $g_2 = \dfrac{GM}{r_2^2}$

$\therefore \dfrac{g_2}{g_1} = \left(\dfrac{r_1}{r_2}\right)^2$. As the data show, these ratios are equal to 2 sf. Hence they provide good support for Newton's law.

[The discrepancy can be attributed to use of data to 3 sf and to rounding.]

Q6 (a) $mr\left(\dfrac{2\pi}{T}\right)^2 = \dfrac{GMm}{r^2}$ or $T = 2\pi\sqrt{\dfrac{r^3}{GM}}$: 1 day = 60 × 60 × 24 s = 86 400 s

$\therefore r^3 = \dfrac{GMT^2}{4\pi^2} = \dfrac{6.67 \times 10^{-11} \times 5.97 \times 10^{24} \times 86\,400^2}{4\pi^2}$

$\therefore r$ = 42 200 000 m

$\therefore$ Height above ground = 42 200 000 − 6 370 000 m = 35 800 km

(b) The orbit must be in the plane of the equator.

Q7   Kepler's 3rd law (for a circular orbit) states that radius$^3 \propto$ period^2

(using the given units) For Phobos: $\dfrac{\text{radius}^3}{\text{period}^2} = \dfrac{9.39^3}{0.319^2}$ = 8140 (3 sf)

For Deimos: $\dfrac{\text{radius}^3}{\text{period}^2} = \dfrac{23.46^3}{1.263^2}$ = 8090 (3 sf)

These values are very close. They are not the same to 3 sf but the 3 sf data are raised to higher powers, which amplifies errors and reduces the number of reliable sf. Hence K3 appears to be reasonably well followed.

Q8 (a) The density at which the universe will slow down asymptotically to zero at infinite expansion.

(b) $[G]$ = N m^2 kg^{-2} = kg m s^{-2} m^2 kg^{-2} = kg^{-1} m^3 s^{-2} ; $[H_0]$ = s^{-1}

$\therefore \left[\dfrac{3H_0^2}{8\pi G}\right] = \dfrac{\text{s}^{-2}}{\text{kg}^{-1}\text{ m}^3\text{ s}^{-2}}$ = kg m^{-3} = $[\rho]$, so dimensionally correct.

Q9 (a) $\dfrac{\Delta\lambda}{\lambda} = \dfrac{v}{c}$, so $v_{max} = \dfrac{3.00 \times 10^8 \times (393.82 - 393.36)}{393.36}$ = 3.51×10^5 m s^{-1} (350 km s^{-1})

$v_{min} = \dfrac{3.00 \times 10^8 \times (393.14 - 393.36)}{393.36}$ = -1.68×10^5 m s^{-1} (−170 km s^{-1})

(b) The star is orbiting a companion. The mean radial velocity is positive, so the system is moving away from the Earth. When the radial velocity is 350 km s^{-1} the star is on the part of its orbit where it is moving away from the Earth. When the radial velocity is −170 km s^{-1}, the star is on the part of its orbit where it is moving towards us, showing that the orbital speed is greater than the recession speed of the system.

Q10 (a) $T = 2\pi\sqrt{\dfrac{d^3}{GM}} = 2\pi\sqrt{\dfrac{(3.0 \times 10^{12})^3}{6.67 \times 10^{-11} \times 4.0 \times 10^{30}}}$ = $2.0(0) \times 10^9$ s

(b) Masses in ratio 3 : 5, so orbit radii in ratio 5 : 3

So Star 1 orbit has radius $\dfrac{5}{8} \times 3.0 \times 10^{12}$ m = 1.9×10^{12} m, and Star 2 orbit has radius 1.1×10^{12} m.

Q11 (a) Peak to peak range of velocities is +350 to –590 m s^{-1}: a range of 940 m s^{-1}

So orbital velocity = $\frac{1}{2}$ × 940 = 470 m s^{-1}.

(b) Period = 3.3 days = 285 000 s

∴ Circumference of orbit = 470 × 285 000 = 1.34 × 10^8 m

∴ Radius = $\dfrac{1.34 \times 10^8}{2\pi}$ = 2.13 × 10^7 m

(c) Period of planet's orbit = 285 000 s

$$T = 2\pi\sqrt{\frac{r^3}{GM}}, \therefore r^3 = \frac{T^2 GM}{4\pi^2} = \frac{(2.85 \times 10^5)^2 \times 6.67 \times 10^{-11} \times 2.6 \times 10^{30}}{4\pi^2}$$

∴ orbital radius, r = 7.1 × 10^9 m

(d) Mass of planet = $\dfrac{2.13 \times 10^7}{(2.13 \times 10^7 + 7.1 \times 10^9)}$ × 2.6 × 10^{30} kg = 7.8 × 10^{27} kg

Q12 (a) Period of orbit, T = 160 day = 1.38 × 10^7 s

Speed of orbit, v = 60 km s^{-1}

∴ $v = \dfrac{2\pi r}{T}, \therefore r_{\text{vis}} = \dfrac{vT}{2\pi} = \dfrac{60 \times 1.38 \times 10^7}{2\pi}$ = 1.32 × 10^8 km

(b) Centripetal force on BH = Gravitational force on BH

∴ $m_{\text{BH}}\left(\dfrac{2\pi}{T}\right)^2 r_{\text{BH}} = \dfrac{Gm_{\text{BH}}m_{\text{vis}}}{(r_{\text{vis}} + r_{\text{BH}})^2}$

Dividing both sides by m_{BH} gives the required equation.

(c) With 8.4 × 10^{10} m:

LHS = $\left(\dfrac{2\pi}{1.38 \times 10^7}\right)^2$ × 8.4 × 10^{10} = 0.0174 m s^{-1}

RHS = $\dfrac{6.67 \times 10^{-11} \times 12 \times 10^{30}}{(2.2 \times 10^{11})^2}$ = 0.0165 m s^{-1} which is the same as LHS to 2 sf.

(d) Using 8.4 × 10^{10} m. m_{BH} × 8.4 × 10^{10} = 12 × 10^{30} × 1.32 × 10^{11}

∴ m_{BH} = 1.9 × 10^{31} kg [~ 10 solar masses]

Q13 (a) H_0 = gradient = $\dfrac{30\,000 \text{ km s}^{-1}}{480 \text{ Mpc}} = \dfrac{3.00 \times 10^7 \text{m s}^{-1}}{480 \times 3.09 \times 10^{22}\text{m}}$ = 2.02 × 10^{-18} s^{-1}

(b) Motion of galaxies within clusters

Q14 If M is the total mass and d the separation: $T = 2\pi\sqrt{\dfrac{d^3}{GM}}$

∴ $M = \dfrac{4\pi^2 d^3}{GT^2} = \dfrac{4\pi^2(30 \times 1.50 \times 10^{11})^3}{6.67 \times 10^{-11} \times (82.2 \times 3.16 \times 10^7)^2}$ = 8.0 × 10^{30} kg = 4$M_\odot$

Radius of orbit of more massive star = $\frac{1}{4}$ of separation.

∴ Mass of less massive star = $\frac{1}{4}$ × 4$M_\odot$ = $M_\odot$; mass of more massive star = 3$M_\odot$

Component 1 Practice paper

Q1 (a) Horizontal distance from A to CoM = $\sqrt{2.00^2 - 1.04^2}$ = 1.708 m

Clockwise moment of weight of ladder = 22.4 × 9.81 × 1.708 = 375.4 N m

∴ for equilibrium 4.00 × F = 375.4

∴ F = 93.8 N ~ 100 N

(b) Horizontal component of F = 93.8 sin θ [where θ = angle between ladder and ground]

$$= 93.8 \times \frac{2.08}{4.00} = 48.8 \text{ N, to the left.}$$

So horizontal component of F_A = 48.8 N to the right.

Vertical component of F = 93.8 × cos θ = 93.8 × 0.854 = 80.1 N

∴ Vertical component of F_A = weight of ladder − 80.1 N

$$= 140 \text{ N upwards}$$

∴ $F_A = \sqrt{140^2 + 48.8^2} = 148$ N at $\tan^{-1}\left(\dfrac{140}{48.8}\right)$ = 70.8° to horizontal (upwards to right)

Q2 (a) The initial velocity, u = 0 and, with a = g and x = h, the equation becomes $h = \frac{1}{2}gt^2$

So, rearranging: $t = \sqrt{\dfrac{2}{g}h} = \sqrt{\dfrac{2}{g}} \times \sqrt{h}$

Comparing this with, y = mx + c, the equation of a straight line of y against x, the gradient, m, is $\sqrt{\dfrac{2}{g}}$ and the intercept is 0 on the t axis.

(b) (i)

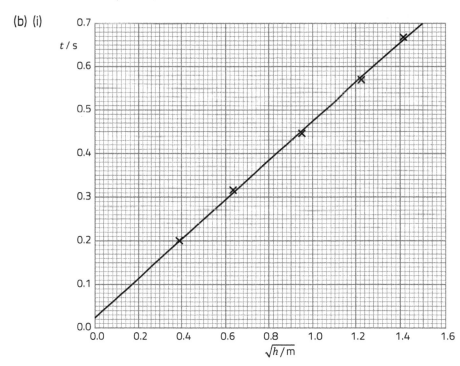

(ii) Gradient = $\dfrac{0.700 - 0.025}{1.50 - 0.00}$ = 0.450

∴ value of g = $\dfrac{2}{0.450^2}$ = 9.88 m s⁻²

(iii) The points lie close to a straight line of best fit, with little scatter which agrees with the theory. However, the graph does not pass through the origin, which the theory predicts (though it is close) and the value of the gradient gives a value for g which is about 1% too high.

(iv) A possible reason is that the magnet does not lose its magnetisation straight away but retains a hold on the ball for about 0.02 s when the switch is opened. There is no reason to alter any of the results because the gradient should still be correct − all the points are just moved upwards by 0.02 s.

Q3 (a) (i) The vector sum of the momenta of the bodies in a system remains constant provided no external resultant force acts.

(ii) Momentum before collision = $0.160 \times 4.33 = 0.693$ N s

Momentum after the collision = $0.160v + 0.160 \times 4.09 = 0.160v + 0.654$ N s

$\therefore 0.160v = 0.693 - 0.654 = 0.039$

$\therefore v = 0.24$ m s^{-1} (2 sf)

(iii) Initial KE = $\frac{1}{2} \times 0.160 \times 4.33^2 = 1.50$ J

Final KE = $\frac{1}{2} \times 0.160 \times 4.09^2 + \frac{1}{2} \times 0.160 \times 0.24^2 = 1.34$ J

Hence 1.50 J − 1.34 J = 0.16 J is lost to the surroundings

(b) Force = momentum change per second.

For the black ball, momentum change = $m\,\Delta v = 0.160 \times 4.09$ N s

$\therefore$ Mean force = $\dfrac{0.160 \times 4.09\,\text{N s}}{16 \times 10^{-6}\,\text{s}} = 41\,000$ N (2 sf) to the right.

The force on the white ball is equal and opposite.

Q4 (a) (i) Initial energy = Initial KE + Initial PE = $0 + 89 \times 9.81 \times 92 = 80\,325$ J

Final energy = Final KE + Final PE = $\frac{1}{2} \times 89 \times 11.0^2 = 5385$ J

$\therefore$ Energy lost due to resistive forces = $80\,352 - 5385 = 75\,000$ J

(ii) This energy is lost to the snowboarder by the work done against air molecules (in drag) and on the snow molecules (in friction). This goes to increasing the internal energy of the atmosphere and snow resulting in a slight rise in temperature and some melting, respectively.

(b) (i) If mean force = F. Work done = $F \times 360$ m = 75 000 J

$\therefore F = 75\,000 / 360$ N = 210 N

(ii) Acceleration $a = \dfrac{v^2}{2x} = \dfrac{11.0^2}{2 \times 360} = 0.168$ m s^{-2}

$\therefore$ Resultant force = $ma = 89 \times 0.168 = 15$ N

Alternatively: Component of weight down slope = $89 \times 9.81 \times 92/360 = 223$ N

$\therefore$ Resultant force = $223 - 210 = 13$ N [nice rounding error!]

(c) In order for thousands of people to spend weeks at high altitude, food must be transported to there, requiring an increase in potential energy of the food, which is achieved by burning fuel in the engines of the transport (or in the power stations which produce electrical energy for the electric vehicles). This increases the load of greenhouse gases in the atmosphere, above that which would have occurred if the skiers had remained at low altitude. They also need heating and need energy for themselves to get there, which also cause environmental damage due to increased CO_2 output.

Q5 (a) (i) & (ii)

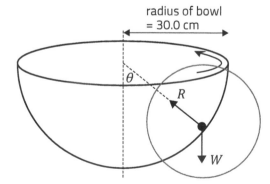

(b) (i) Centripetal force is the resultant force on an object, which is travelling in a circle at a constant speed. It is directed towards the centre of the circle.

(ii) The horizontal component of $R = R\sin\theta$ and this provides the centripetal force for the horizontal circular motion of the ball.

Hence $R\sin\theta = mr\omega^2$

(iii) There is no motion in the vertical direction, so the resultant vertical component of force is zero.

Vertical component of $R = R\cos\theta$ and $W = mg$. Hence, $R\cos\theta = mg$

(iv) From the two equations: $\dfrac{R\sin\theta}{R\cos\theta} = \dfrac{mr\omega^2}{mg}$, $\therefore \tan\theta = \dfrac{r\omega^2}{g}$

$\theta = \sin^{-1}\left(\dfrac{15}{30}\right) = 30°$, $\therefore \tan\theta = \dfrac{1}{\sqrt{3}}$, $\therefore \omega^2 = \dfrac{9.81 \times 1}{\sqrt{3} \times 0.15}$, $\therefore \omega = 6.15$ rad s^{-1}

$\therefore$ period $T = \dfrac{2\pi}{\omega} = \dfrac{2\pi}{6.15} = 1.02$ s so Jasmine is correct.

Q6 (a) $mg = k\Delta x$

$\therefore \Delta x = \dfrac{45 \times 9.81}{5320} = 0.083$ m

(b) $T = 2\pi\sqrt{\dfrac{m}{k}} = 2\pi\sqrt{\dfrac{45}{5320}} = 0.578$ s

$\therefore$ 5 oscillations should take 5×0.578 s $= 2.89$ s, which is not consistent.

The chair and the boy both oscillate so the total mass will be significantly greater than 45 kg, leading to a greater period.

(Another possible answer is that she miscounted the number of oscillations. If there were 6 oscillations in 3.55 s, this would give a period of 0.59 s, which is pretty close to the theoretical value.)

(c) At the top of the motion all the energy is a combination of gravitational potential energy and elastic potential energy.

As the chair moves towards the centre of the motion, the gravitational potential energy decreases and the elastic potential energy and kinetic energy both increase.

In the second half of the downwards motion the gravitational potential energy continues to decrease; the kinetic energy also decreases but the elastic potential energy increases.

Q7 (a) $pV = nRT$, so $n = \dfrac{pV}{RT} = \dfrac{50\,000 \times 2.4}{8.31 \times 293} = 49.3$ mol

(b) Pressure at B is half pressure at A but the volume is the same. So the temperature is half that at A, i.e. 146.5 K

Volume at C is $3.64/2.40 \times$ that at B, the pressure is the same, so the temperature is 222 K

(c) AB: $W = 0$ because the volume is constant.

BC: W = area under the graph $= p\Delta V = 25\,000 \times (3.64 - 2.40) = 31\,000$ J

CA: Area under graph $= 31\,000 +$ area enclosed $= 31\,000 + 12 \times 5\,000 \times 0.2$
$\qquad\qquad\qquad\qquad = 43\,000$ J

This is the work done **on** the gas, so work done **by** gas $= -43\,000$ J.

Cycle: $0 + 31\,000 - 43\,000 = -12\,000$ J

(d) Internal energy increase $= \dfrac{3}{2}nR\Delta T = \dfrac{3}{2} \times 49.3 \times 8.31 \times 71 = 43\,600$ J

The increase in internal energy in CD is almost the same as the work done by the gas so there is little or no heat transfer, i.e. David is correct (to within the uncertainty of the data, especially estimating the area under the curve).

Q8 **Assumptions**

- Gases consist of a large number of molecules in rapid random motion.
- The space occupied by the molecules is a negligible fraction of the volume of the gas.
- The molecules behave as perfectly elastic spheres.
- The molecules exert negligible forces on each other except during collisions.

How pressure arises

A large number of molecules collide with any small area of the container wall per second, rebounding back into the gas. Thus the gas molecules suffer a change in momentum at the wall. The force exerted by the wall on the molecules is the change of momentum of the molecules per second. By Newton's third law, the molecules exert an equal and opposite force on the wall of the container. The pressure is the magnitude of this force per unit area of wall.

Component 2 Practice paper

Q1 (a) In the steep linear part, increasing the tension force increases the bond length between atoms (in the direction of the tension). This is the elastic region.

At the top of the straight part, at the yield point, edge dislocations become mobile: bonds at the end of crystal half-planes break and reform with the half-plane moving in the direction of the stress. At slightly higher stress, new edge dislocations are formed at the site of weaknesses, e.g. missing or additional impurity atoms, and also move. This movement of dislocations produces plastic deformation.

With increasing elongation a neck is produced where more dislocations have moved, causing a higher true stress, causing further movement of dislocations until the material breaks.

(b) If T is the tension in each wire, $2T \sin 20° = 5.5 \times 9.81$

$\therefore T = 78.9$ N

Stress, $\sigma = \dfrac{T}{\pi r^2} = \dfrac{78.9 \text{ N}}{\pi \times (0.18 \times 10^{-3} \text{ m})^2} = 775$ MPa

This is lower than the breaking stress but it is very close, so in engineering terms the wires are not appropriate. A factor of safety of about 5 is normally applied, suggesting that double this diameter would be appropriate.

Q2 (a) (i) Drift velocity is given by $v = \dfrac{I}{nAe}$, where n is the number of charge carriers per unit volume. Because the materials are the same, n will be the same. The current is the same in the two wires (because they are in series).

So $v \propto \dfrac{1}{A} \propto \dfrac{1}{\text{diameter}^2}$ Hence $v_Y = 9 \times v_X$, i.e. ratio = 9

(ii) Assuming that the material of the two wires is the same, the resistivity r is the same.

$P = I^2 R$. The current is the same, so $P \propto R \propto \dfrac{l}{A}$

$\therefore \dfrac{P_Y}{P_X} = \dfrac{3/\frac{1}{9}}{1/1} = 27$

(b) (i) Given the small uncertainty, the measuring instrument would be a digital calliper.

(ii) Take measurements in pairs at right angles; make sure the wire is straight (free from kinks); make sure that the callipers are zeroed (or take a zero reading).

(iii) $R = \dfrac{\rho L}{A} = \dfrac{\rho L}{\pi r^2} = \dfrac{4\rho L}{\pi d^2} = \dfrac{4\rho L}{\pi} \times \dfrac{1}{d^2}$

Comparing with $y = mx + c$, the gradient is $\dfrac{4\rho L}{\pi}$.

(iv) Max gradient $= \dfrac{18.0 - 0.0}{(27.0 - 0.9) \times 10^6} = 6.90 \times 10^{-7}$ Ω m²

Min gradient $= \dfrac{17.25 - 0.50}{(30.0 - 0.0) \times 10^6} = 5.58 \times 10^{-7}$ Ω m2

Gradient $= (6.24 \pm 0.66) \times 10^{-7}$ Ω m: fractional uncertainty = 0.105

$\rho = $ gradient $\times \dfrac{\pi}{4L} = 1.96 \times 10^{-7}$ Ω m: uncertainty 0.21

So $\rho = (2.0 \pm 0.2) \times 10^{-7}$ Ω m

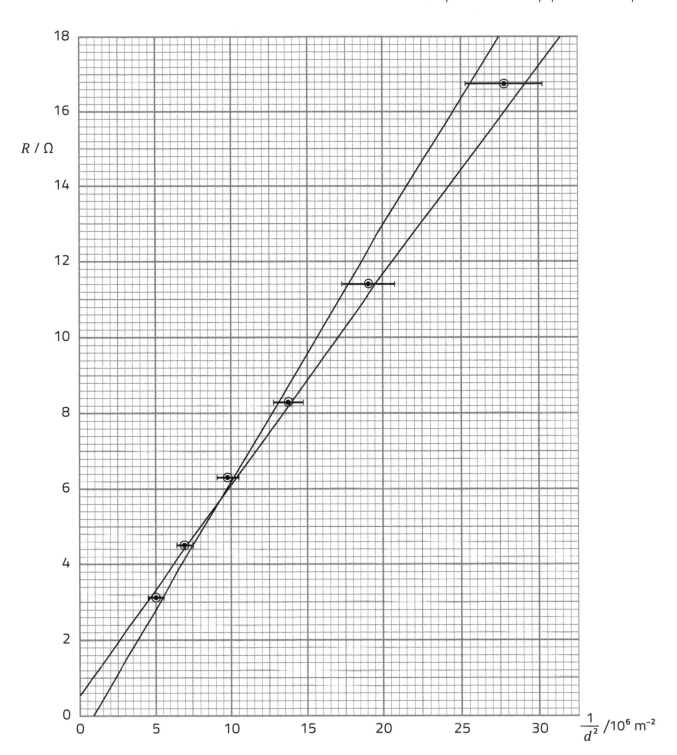

Q3 (a) The capacitance of a capacitor is the charge transferred from one plate to the other per unit pd.

(b) (i) Either: $C = \dfrac{\varepsilon_0 A}{d}$, so, for $C = 3$ nF:

$$d = \frac{\varepsilon_0 A}{C} = \frac{8.85 \times 10^{-12} \times (0.25)^2}{3 \times 10^{-9}} = 1.84 \times 10^{-4} \text{ m} \sim 0.2 \text{ mm}$$

Or, with a gap of 0.2 mm,

$$C = \frac{\varepsilon_0 A}{d} = \frac{8.85 \times 10^{-12} \times (0.25)^2}{0.2 \times 10^{-3}} = 2.76 \times 10^{-9} \text{ F} \sim 3 \text{ nF}$$

(ii) $V = \dfrac{Q}{C}$ and $E = \dfrac{V}{d}$, so $E = \dfrac{Q}{Cd} = \dfrac{60 \text{ nC}}{3 \text{ nF} \times 1.84 \times 10^{-4}\text{m}} = 110 \text{ kV m}^{-1}$ [or N C^{-1}]

[or 100 kV m^{-1} if $d = 0.2$ mm used]

(iii) Time constant, $\tau = \frac{1}{RC}$. So to obtain 30 s, $R = \frac{1}{30 \times 3 \times 10^{-9}} = 11 \, M\Omega$

This is a very high resistance; and circuit to which it is connected is likely to have a resistance lower than this, which will change to time constant.

(c) (i) Total charge = 12 × 10 = 120 mC

New total capacitance = 10 + 15 = 25 mF

∴ New pd $= \frac{Q}{C} = \frac{120 \, mC}{25 \, mF} = 4.8$ V

(ii) $W = \frac{1}{2}QV = \frac{1}{2} \times 120 \, mC \times 4.8$ V = 290 mJ (2 sf)

(iii) Rapid oscillations in the current in the circuit at the moment of connection produce radio waves which carry away energy. A spark is produced which also radiates light. There is also a heating effect in the wires connecting the two capacitors.

Q4 (a) (i) Pd across 30 Ω = 4 V, ∴ $P = \frac{4^2}{30} = 0.53$ W, so 1.0 W is adequate.

Power dissipated by 60 Ω resistor = 2 × 0.53 = 1.1 W. So 1.0 W is inadequate.

(ii) Without 180 Ω resistor, $V_{out} = 8.0$ V

With 180 Ω resistor, output resistance = 45 Ω

∴ $V_{out} = 3/5 \times 12$ V = 7.2 V

∴ $\Delta V_{out} = -0.8$ V

(b) In the daylight, the LDR resistance will be lower, so the fraction of the pd across R_0 is greater, i.e. V_{out} is greater and so Paul is correct.

If R_0 is increased the fraction of the pd across it will decrease so V_{out} will decrease and Paul is incorrect.

Q5 (a)

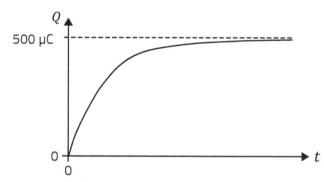

(b) $Q = Q_0 (1 - e^{-t/RC})$, ∴ If $Q = 0.5Q_0$, $e^{-t/RC} = 0.5$ ∴ $e^{-t/RC} = 2$

∴ $\frac{t}{RC} = \ln 2$, ∴ $t = 150 \, k\Omega \times 50 \, mF \times \ln 2 = 5.2$ s (2 sf)

(c) (i) The current is the rate of change of charge on the capacitor, which is the gradient of the graph, which decreases with time.

(ii) The current causes the charges (±Q) on the capacitor plates to increase. This increases the pd across the capacitor (V = Q/C). Hence the pd across the series resistor decreases (they add up to 10 V). If the pd across the resistor decreases, so does the current, by Ohm's law. So Trisha is correct.

Q6 (a) E_p due to each charge $= \frac{Q}{4\pi\varepsilon_0 d^2} = 9 \times 10^9 \frac{12 \times 10^{-12}}{(50 \times 10^{-3})^2} = 43.2 \, V \, m^{-1}$

Horizontal component = 4/5 of this [or use cos 37°]. The horizontal components add and the vertical components cancel.

∴ $E_p = 1.6 \times 43.2 = 69 \, V \, m^{-1}$ to the right.

(b) At O, the electric fields due to the two charges are equal and opposite, so the resultant field is zero (as shown in the graph).

As $x \longrightarrow \infty$, the fields from each charge $\longrightarrow 0$, so the resultant field tends to zero as shown.

The resultant field is always to the right between 0 and ∞ and has a non-zero value. As it is 0 when $x = 0$, zero as $x \rightarrow \infty$ and positive between, the shape of the graph is reasonable.

(c) (i) This is the work done per unit charge in bringing a small test charge from infinity [or another point of defined zero potential] to that point.

(ii) The potential at O = $2 \times \dfrac{1}{4\pi\varepsilon_0} \times \dfrac{12 \times 10^{-12}}{0.030}$ V = 7.2 V

∴ Initial potential energy = 7.2 V × (−3.2 × 10^{-19} C) = −2.30 × 10^{-18} J

The initial kinetic energy = $\frac{1}{2}$ × 9.35 × 10^{-26} kg × (5000 m s^{-1})^2 = 1.17 × 10^{-18} J

∴ Total energy at O = (1.18 − 2.30) × 10^{-18} J = −1.12 × 10^{-18} J

The potential at P = $2 \times \dfrac{1}{4\pi\varepsilon_0} \times \dfrac{12 \times 10^{-12}}{0.050}$ = 4.32 V

∴ Potential energy at P = 4.32 V × (−3.2 × 10^{-19} C) = −1.38 × 10^{-18} J

This is less than the total energy at O, so the particle does pass point P [with a kinetic energy of 0.26 × 10^{-18} J]

Q7 (a) λ_{max} = 620 nm

So, assuming that the star emits as a black body, $T = \dfrac{2.90 \times 10^{-3}\,\text{m K}}{620 \times 10^{-9}\,\text{m}}$ = 4680 K

(b) (i) The intensity, I, of the radiation received on Earth is related to the power, L, emitted by the star by the equation

$$I = \frac{L}{4\pi r_{PE}^2}$$

So, if the received intensity is measured, L can be calculated.

(ii) $L = A\sigma T^4$, so $4\pi \left(\dfrac{d}{2}\right)^2 = \dfrac{L}{\sigma T^4} = \dfrac{1.65 \times 10^{28}}{5.67 \times 10^{-8} \times 4680^4}$ = 6.076 × 10^{20} m^2

∴ d = 1.4 × 10^{10} m

(c) The hot atmosphere of Pollux consists of a tenuous gas of atoms (many ionised). The star emits photons with a continuous range of frequencies (i.e. energies). They can absorb photons which have an energy equal to differences in the energy levels of the atoms – promoting the atoms to higher energy levels. The atoms subsequently lose energy by emitting photons in random directions, so there are fewer photons with these specific energies – hence dark lines in the spectrum.

Q8 (a) $\dfrac{\Delta\lambda}{\lambda} = \dfrac{v}{c}$, ∴ $\Delta\lambda = \dfrac{v\lambda}{c} = \dfrac{0.3 \times 397 \times 10^{-9}}{3.00 \times 10^8}$ = 4.0 nm

(b) (i) Orbital speed = 47 m s^{-1}; period = 590 days = 5.10 × 10^7 s

∴ Circumference = 2.40 × 10^9 m

∴ Radius = $\dfrac{2.40 \times 10^9}{2\pi}$ = 3.81 × 10^8 m

(ii) Ignoring the mass of the planet, i.e. assuming it to be much less than the mass of β Gem:

$T = 2\pi\sqrt{\dfrac{d^3}{GM}}$, so $d^3 = \dfrac{T^2 GM}{4\pi^2} = \dfrac{(5.10 \times 10^7)^2 \times 6.67 \times 10^{-11} \times 3.8 \times 10^{30}}{4\pi^2}$

∴ d = 2.55 × 10^{11} m.

(iii) Mass of planet = $\dfrac{3.81 \times 10^8\,\text{m}}{2.55 \times 10^{11}\,\text{m}} \times 3.8 \times 10^{30}$ kg = 5.7 × 10^{27} kg

(c) Many areas of pure science, e.g. the work of J. J. Thomson leading to the discovery of the electron, have had many practical spin-offs eventually. These were not anticipated at the time. Likewise, with electromagnetic induction. Hence this is a point of view which can be challenged. Astronomical discoveries generally fall into the category of blue-sky [well, black-sky] research, whose only purpose is to advance knowledge of the universe. The search for potential life-bearing exoplanets has a big bearing on the position of life on Earth (and therefore human life) vis-a-vis the universe ... .
[Note: You wouldn't be expected to write as much. This answer gives some idea of the sort of points which the examiner would hope to see.]